Pun

Numerical Simulations of the Head-Disk Interface in Hard Disk Drives

Puneet Bhargava

Numerical Simulations of the Head-Disk Interface in Hard Disk Drives

Finite Element Solutions for the Transient and Steady State Air-Bearing Equations

VDM Verlag Dr. Müller

Impressum/Imprint (nur für Deutschland/ only for Germany)
Bibliografische Information der Deutschen Nationalbibliothek: Die Deutsche Nationalbibliothek verzeichnet diese Publikation in der Deutschen Nationalbibliografie; detaillierte bibliografische Daten sind im Internet über http://dnb.d-nb.de abrufbar.
Alle in diesem Buch genannten Marken und Produktnamen unterliegen warenzeichen-, marken- oder patentrechtlichem Schutz bzw. sind Warenzeichen oder eingetragene Warenzeichen der jeweiligen Inhaber. Die Wiedergabe von Marken, Produktnamen, Gebrauchsnamen, Handelsnamen, Warenbezeichnungen u.s.w. in diesem Werk berechtigt auch ohne besondere Kennzeichnung nicht zu der Annahme, dass solche Namen im Sinne der Warenzeichen- und Markenschutzgesetzgebung als frei zu betrachten wären und daher von jedermann benutzt werden dürften.

Coverbild: www.purestockx.com

Verlag: VDM Verlag Dr. Müller Aktiengesellschaft & Co. KG
Dudweiler Landstr. 99, 66123 Saarbrücken, Deutschland
Telefon +49 681 9100-698, Telefax +49 681 9100-988, Email: info@vdm-verlag.de
Zugl.: Berkeley, University of California, 2008

Herstellung in Deutschland:
Schaltungsdienst Lange o.H.G., Berlin
Books on Demand GmbH, Norderstedt
Reha GmbH, Saarbrücken
Amazon Distribution GmbH, Leipzig
ISBN: 978-3-639-13214-4

Imprint (only for USA, GB)
Bibliographic information published by the Deutsche Nationalbibliothek: The Deutsche Nationalbibliothek lists this publication in the Deutsche Nationalbibliografie; detailed bibliographic data are available in the Internet at http://dnb.d-nb.de.
Any brand names and product names mentioned in this book are subject to trademark, brand or patent protection and are trademarks or registered trademarks of their respective holders. The use of brand names, product names, common names, trade names, product descriptions etc. even without a particular marking in this works is in no way to be construed to mean that such names may be regarded as unrestricted in respect of trademark and brand protection legislation and could thus be used by anyone.

Cover image: www.purestockx.com

Publisher:
VDM Verlag Dr. Müller Aktiengesellschaft & Co. KG
Dudweiler Landstr. 99, 66123 Saarbrücken, Germany
Phone +49 681 9100-698, Fax +49 681 9100-988, Email: info@vdm-publishing.com
Berkeley, University of California, 2008

Printed in the U.S.A.
Printed in the U.K. by (see last page)
ISBN: 978-3-639-13214-4

To my Parents

Contents

List of Figures   **vii**

List of Tables   **xi**

Acknowledgements   **xii**

Symbols and Abbreviations   **xiv**

1 Introduction   **1**
    1.1   Basics 1
    1.2   Evolution 3
    1.3   Motivation 4
    1.4   Outline 5
    1.5   Figures 6

2 Structural modeling   **10**
    2.1   Introduction 10
    2.2   The HGA model 11
    2.3   Disk model 12
    2.4   Governing Equation 13
       2.4.1   Linearization 14
    2.5   Guyan reduction 15
    2.6   Modeling suspension nonlinearities 16
    2.7   Time discretization 17
    2.8   Discussion 17
    2.9   Figures 18

3 Contact modeling   **21**
    3.1   Introduction 21
    3.2   Asperity Contact forces 22
    3.3   Impact forces 22
       3.3.1   Assumptions 22

	3.3.2	Formulation	23
	3.3.3	The Boussinesq problem	24
	3.3.4	The Boundary Element Method	24
	3.3.5	Calculation of impact stiffness	27
	3.3.6	Radius of effect	29
	3.3.7	Solution comparison	29
3.4	Discussion		31
3.5	Figures		32

4 Fluid static modeling   **39**
4.1	Introduction		39
4.2	Previous Work		40
4.3	Methodology		42
	4.3.1	Governing equations	42
	4.3.2	The Weak Form	44
	4.3.3	The Forward Problem	45
	4.3.4	Inverse problem	49
	4.3.5	Inverse Problem Method 1: Series of Forward Problems	49
	4.3.6	Inverse Problem Method 2: Fully linearized system	52
	4.3.7	Streamline Upwind/Petrov-Galerkin formulation	56
	4.3.8	The force norm	57
4.4	Numerical Simulations		57
	4.4.1	Meshing	57
	4.4.2	Implementation	58
	4.4.3	Forward problem	59
	4.4.4	Refinement strategies	60
	4.4.5	Inverse problem	65
4.5	Discussion		65
4.6	Figures		67

5 Fluid dynamic modeling   **96**
5.1	Introduction		96
5.2	Previous Work		96
5.3	Methodology		97
	5.3.1	The generalized Reynolds Equation	97
	5.3.2	The Weak Form	99
	5.3.3	Discretization in time	100
	5.3.4	Linearization	100
	5.3.5	Streamline Upwind/Petrov-Galerkin formulation	101
	5.3.6	Finite element discretization	101
	5.3.7	Mesh generation	103
	5.3.8	Assembly	103
	5.3.9	Solution	103
	5.3.10	Air bearing damping/stiffness calculation	104

      5.3.11 Algorithmic air bearing stiffness 107

5.4 Numerical Simulations 109

5.5 Discussion 109

5.6 Figures 110

6 Simulation of the shock event   **114**

6.1 Introduction 114

6.2 Prior Work 115

6.3 The CML L/UL/S Simulator 116

      6.3.1 Suspension modeling 117

      6.3.2 Air bearing modeling 117

      6.3.3 Coupling 117

6.4 Methodology 118

      6.4.1 Modeling Shock 118

6.5 Pulse-width effect 118

      6.5.1 Component Dynamics 119

      6.5.2 Results and Discussion 121

      6.5.3 Conclusions 123

6.6 Effect of dimple location 124

      6.6.1 Procedure 124

      6.6.2 Results and Discussion 125

      6.6.3 Conclusions 129

6.7 Discussion 130

6.8 Tables 131

6.9 Figures 132

7 Simulating the Load/Unload Process   **159**

7.1 Introduction 159

7.2 Previous work 161

7.3 Methodology 162

7.4 Numerical Simulations 162

      7.4.1 Unloading process 163

      7.4.2 Loading process 165

      7.4.3 Frequency response 167

7.5 Conclusion 168

7.6 Figures 169

8 Conclusion and Future Work   **183**

8.1 Conclusion 183

8.2 Future Work 186

A The CML Dynamic Load/Unload/Shock Simulator (Version 5.1)   **197**

  A.1   Introduction 197

  A.2   Installation 199

  A.3   Procedure 199

  A.4   Suspension model 200

  A.5   Input files 202

      A.5.1   rail.dat 202

      A.5.2   dynamics.def 203

      A.5.3   mass.txt, stiffness.txt, coords.txt 207

  A.6   Output files 207

      A.6.1   attitude.dat 208

      A.6.2   cestatii.dat 208

      A.6.3   cpressures.dat 208

      A.6.4   imf.dat 208

      A.6.5   lultab.dat 208

      A.6.6   shock.dat 208

  A.7   Post-processing 209

      A.7.1   displaceLUL.m 209

      A.7.2   displaceSHOCK.m 209

      A.7.3   force.m 210

      A.7.4   cestat.m 210

      A.7.5   lulbehav.m 211

      A.7.6   actmot.m 211

  A.8   Figures 212

List of Figures

1.1 A typical modern hard disk drive 6
1.2 The IBM 350 Disk Storage system 7
1.3 The Hitachi Deskstar 7K1000 7
1.4 Evolution of areal density in hard disk drives 8
1.5 Various layers on the disk in a HDD 9
1.6 The head gimbal assembly 9

2.1 An HSA model for a popular 1" drive 18
2.2 Schematic of the HSA 18
2.3 Schematic of the HSA showing limiters engaged 18
2.4 First axisymmetric (umbrella) mode of the disk 19
2.5 First radial mode of the disk 19
2.6 First axisymmetric-radial coupled mode 19
2.7 Disk response to z-shock for spinning and stationary disks 20
2.8 Power spectra of disk shock response for spinning and stationary disks . . . 20

3.1 The Boussinesq problem 32
3.2 Slider disk contact 32
3.3 Elemental basis functions 33
3.4 Reference element 33
3.5 Convergence as a function of 'Radius of Effect' 34
3.6 Radial pressure profile for a rigid flat-end circular indenter 34
3.7 Radial pressure profile for a rigid parabolic circular indenter 35
3.8 Total contact force for a rigid parabolic circular indenter 35
3.9 Slider profile 36
3.10 Finite element mesh 36
3.11 Pressure profile during Slider-disk contact 1 37
3.12 Pressure profile during Slider-disk contact 2 37
3.13 Pressure profile during Slider-disk contact 3 38

4.1 Mesh Coordinate System 67
4.2 System Coordinate System 67

viii

4.3 Basis Functions 68
4.4 Coarse level system matrix before node reordering for the forward problem 69
4.5 Coarse level system matrix after RCM ordering for Method 1 69
4.6 Coarse level system matrix after RCM ordering for Method 2 70
4.7 Pressure solution for a 2-rail taper slider with and without upwinding . . . 71
4.8 Element length scale 72
4.9 Coarse level meshes for two slider designs 72
4.10 Flat slider profile 73
4.11 2-rail slider 73
4.12 Slider 1 profile 73
4.13 Grid convergence for flat slider 74
4.14 Force error for flat slider 74
4.15 Converged pressure profile for flat slider 75
4.16 Grid convergence for 2-rail slider 76
4.17 Force error for 2-rail slider 76
4.18 Converged pressure profile for 2-rail slider 77
4.19 Grid convergence for slider 1 78
4.20 Force error for slider 1 78
4.21 Converged pressure profile for slider 1 79
4.22 Slider design used for refinement studies 80
4.23 Refinement slider pressure profile 80
4.24 Coarse mesh error for refinement slider 81
4.25 Coarse mesh for refinement slider 81
4.26 Estimated solution error using pressure based mesh refinement 82
4.27 Refined mesh generated using pressure based mesh refinement 82
4.28 Estimated solution error using clearance based mesh refinement 83
4.29 Refined mesh generated using clearance based mesh refinement 83
4.30 Estimated solution error using pressure gradient based mesh refinement . . 84
4.31 Refined mesh generated using pressure gradient based mesh refinement . . . 84
4.32 Estimated solution error using flux jump based mesh refinement 85
4.33 Refined mesh generated using flux jump based mesh refinement 85
4.34 Estimated solution error using clearance gradient based mesh refinement . . 86
4.35 Refined mesh generated using clearance gradient based mesh refinement . . 86
4.36 Estimated solution error using clearance gradient based mesh refinement . . 87
4.37 Refined mesh generated using RE residual based mesh refinement 87
4.38 Grid convergence for refinement slider using various refinement strategies . 88
4.39 Grid convergence for slider 1 using various refinement strategies 89
4.40 Grid convergence for slider 2 using various refinement strategies 90
4.41 Slider 2 profile 91
4.42 Slider 3 profile 91
4.43 Grid convergence for slider 1 92
4.44 Mesh levels for slider 1 93
4.45 Grid convergence for slider 2 94
4.46 Grid convergence for slider 3 95

5.1 System Coordinate System 110
5.2 Slider 2 110
5.3 Slider attitude for free vibration simulation 111
5.4 Slider forces for free vibration simulation 112
5.5 Error-estimate/time-step for free vibration simulation 113

6.1 Half-sine shock pulse 132
6.2 The disk umbrella mode 132
6.3 Disk response for 200G shock of varying pulse widths 133
6.4 Maximum disk deflection for 200G shock of varying pulse widths 133
6.5 Disk response spectra for 200G shock of varying pulse widths 134
6.6 Suspension schematic 134
6.7 Suspension first bending mode 135
6.8 Suspension first torsion mode 135
6.9 Suspension second bending mode 135
6.10 Suspension response to a 200G shock of varying pulse widths 136
6.11 Suspension response spectra for a 200G shock of varying pulse widths . . . 136
6.12 Structural Air-bearing coupling scheme 137
6.13 Slider Design 138
6.14 Slider Response for 125G, 0.2ms shock 139
6.15 Frequency spectra of load beam and disk response to 125G, 0.2ms shock . . 140
6.16 Dimple, limiter status for 125G, 0.2ms shock 141
6.17 Air bearing and contact forces for 125G, 0.2ms shock 142
6.18 Slider Response for 250G, 0.5ms shock 143
6.19 Slider Response for 375G, 1.0 ms shock 144
6.20 Slider Response for 400G, 1.0 ms shock 145
6.21 Dimple, limiter status for 400G, 1.0 ms shock 146
6.22 Slider Response for 400G, 2.0 ms shock 147
6.23 Slider Response for 450G, 3.0 ms shock 148
6.24 Safe shock levels for varying pulse widths 149
6.25 Suspension system 150
6.26 Slider free-body diagram 150
6.27 Slider attitude for BC 400G shock 151
6.28 Dimple separation and contact force for BC 400G shock 152
6.29 Slider attitude for U1 400G shock 153
6.30 Dimple separation and contact force for U1 400G shock 154
6.31 Linearized system 154
6.32 $c_{z_{min}G}$ variation along x_l and y_l 155
6.33 Slider attitude for U2 400G shock 156
6.34 Dimple separation and contact force for U2 400G shock 157
6.35 Safe shock levels for various dimple locations 158

7.1 Schematic of a 1" drive 169
7.2 Slider used for simulations 169

7.3   Slider attitude history comparison during the unloading process (44.4 mm/s) 170

7.4   Force history comparison during the unloading process (44.4 mm/s)   171

7.5   Ramp-tab position and ramp contact force during the loading process (44.4
      mm/s)   172

7.6   Dimple and limiter contact status during the loading process (44.4 mm/s) .   172

7.7   Slider attitude history comparison during the loading process (88.9 mm/s) .   173

7.8   Force history comparison during the unloading process (88.9 mm/s)   174

7.9   Ramp-tab position and ramp contact force during the loading process (88.9
      mm/s)   175

7.10 Dimple and limiter contact status during the loading process (88.9 mm/s) .   175

7.11 Slider attitude history comparison during the loading process (50 mm/s) . .   176

7.12 Force history comparison during the loading process (50 mm/s)   177

7.13 Ramp-tab position and ramp contact force during the loading process (50
      mm/s)   178

7.14 Dimple and limiter contact status during the loading process (50 mm/s) . .   178

7.15 Slider attitude history comparison during the loading process (66.7 mm/s) .   179

7.16 Force history comparison during the loading process (66.7 mm/s)   180

7.17 Ramp-tab position and ramp contact force during the loading process (66.7
      mm/s)   181

7.18 Dimple and limiter contact status during the loading process (66.7 mm/s) .   181

7.19 Frequency spectra of a typical L/UL process for 4-DOF (a, b, c) and FE
      based (d, e, f) simulators   182

A.1   Actuator nomenclature   212

A.2   The suspension coordinate system   213

A.3   The slider coordinate system   213

A.4   The ramp coordinates   214

A.5   Sample ramp profile   214

A.6   Slider attitude for example 1 plotted using the *displaceLUL* command . . .   215

A.7   Slider attitude for example 3 plotted using the *displaceSHOCK* command .   216

A.8   Slider attitude for example 4 plotted using the *displaceSHOCK* command .   217

A.9   Air bearing and contact forces for example 2 plotted using the *force* command 218

A.10 Air bearing and contact forces for example 3 plotted using the *force* command 219

A.11 Contact element data for example 2 plotted using the *cestat* command . . .   220

A.12 Contact element data for example 3 plotted using the *cestat* command . . .   221

A.13 L/UL tab behavior for example 1 plotted using the *lulbehav* command . . .   222

A.14 Actuator motion for example 1 plotted using the *actmot* command   223

List of Tables

4.1   Rarefaction and slip models   43
4.2   Slider attitudes for grid convergence studies   60
4.3   Slider parameters for mesh refinement study   61
4.4   Slider operation parameters for grid convergence studies   66

6.1   Disk parameters for shock studies   131
6.2   Slider parameters for shock studies   131

7.1   Slider parameters for load/unload comparison studies   163

Acknowledgements

First and foremost I would like to thank my advisor and mentor, Prof. David Bogy for his guidance these past five years, not only in areas of research but also life in general. All through my research career he has given me the freedom to explore research areas of my interest while at the same time not letting me stray off course. It has been an absolute honor to work with him during my graduate studies here at Berkeley..

I owe a great deal to Dr. C. Singh Bhatia for his guidance and for his insightful feedback at our weekly group meetings. I would also like to thank Professor Tarek Zohdi and Professor John Strain for their valuable feedback while reviewing this dissertation. During my CML years, I have also had the opportunity of close interaction with industry. I would like to thank Samuel Gan of Seagate Singapore for the helpful discussions and for supporting part of this project and providing models and equipment for research.

A big thanks is also due to fellow CMLers. Technical discussions with them have helped me gain a broader education. Brendan Cox, Sujit Kirpekar, Rohit Ambekar, Vineet Gupta, Sripathi Canchi and Rahul Rai deserve special thanks. Life in Berkeley would not have been fun without these wonderful friends: Ankit Jain, Ambuj Tewari, Anurag Gupta and Vishnu Narayanan.

Finally, I attribute all my success to my wonderful family. I am indebted to my them for their sacrifices and unconditional love throughout my life. Mere words cannot express my gratitude to them.

The research presented in this dissertation was supported by the Computer Mechanics Laboratory at the University of California, Berkeley, and Seagate Singapore.

Symbols

Variables

P — Nondimensional normalized pressure

p — Pressure

p_0 — Ambient pressure

H — Nondimensional clearance

h — Clearance

h_m — Clearance length scale

T — Nondimensional time

t — Time

ω — Angular velocity of the disk

$\mathcal{S}$ — Domain of the air bearing slider

$\partial\mathcal{S}$ — Boundary of the air bearing slider

X, Y — Nondimensional coordinates

x, y — Dimensional coordinates

L — Characteristic slider length

$\mathbf{\Lambda}$ — Bearing number

μ — Dynamic viscosity of air

$\mathbf{U}$ — Local disk velocity

τ — Squeeze number

v — Test function

∂P Update to normalized pressure P

P_i Nodal normalized pressure

ϕ_i Basis for clearance and pressure

$\tilde{\phi}_i$ Basis for test functions

Abbreviations

ABS   Air-Bearing Surface

BEM   Boundary Element Method

CML   Computer Mechanics Laboratory

CSS   Contact Start/Stop

DLC   Diamond like Carbon

DOF   Degree of Freedom

FEM   Finite Element Method

FH   Fly-Height

HDD   Hard Disk Drive

HDI   Head Disk Interface

HAA   Head Actuator Assembly

HGA   Head Gimbal Assembly

HSA   Head Stack Assembly

ID   Inner Diameter

L/UL Load/Unload

MD   Middle Diameter

MFP   Mean Free Path

OD   Outer Diameter

PC   Personal Computer

PSA   Pitch Static Attitude

RMS   Root Mean Square

RPM   Revolutions Per Minute

RSA   Roll Static Attitude

SUPG Streamline Upwind Petrov Galerkin

TEC   Trailing Edge Center

Chapter 1

Introduction

A hard disk drive is one of the most important and also one of the most interesting components within the PC. They have a long and interesting history dating back to the early 1950s. A hard disk drive (HDD), is a non-volatile storage device which stores digital data on spinning disks with magnetic surfaces. The research presented in this dissertation was carried out at the Computer Mechanics Laboratory (CML) at the University of California, Berkeley and is focused on the numerical simulations of the dynamics of the HDD. In the following sections we present an introduction to the mechanical workings of the HDD, followed by a brief history of the evolution of HDDs. We also present the motivation for this research and then present a brief outline of how this dissertation is organized.

1.1 Basics

A picture of a modern hard disk drive is shown in Fig. 1.1 [Source: Western Digital Corporation, 2008]. The main components of a typical hard disk drive are the spindle with the disk stack (platters), the actuator and the actuator arm and the logic board. Information

is written on (and read back from) concentric *tracks* on each platter by the read/write

transducer which are located at the trailing edge of a *slider*. The slider is mounted on a

suspension (together known as the head-gimbal assembly, or the HGA) many of which are in

turn mounted on the e-block arm (one for each data surface). The HGA-Eblock assembly is

known as the head stack assembly (HSA). The HSA rotates about the actuator axis, and is

thus able to access all the tracks on the disk. The disk stack typically rotates at 3600-15000

RPM and this creates an air-bearing under the surface of the slider, which causes it to lift

off and 'float' a few nanometers above the surface of the disk.

The disks used in the drives are ultra-smooth (RMS roughness ∼0.2 nm) and are

typically made of glass or aluminum substrate with multiple layers (see Fig. 1.5, [Source: UK

Data Recovery and HDD repair, 2008]). The data is stored as the magnetic polarization of

bits on the magnetic layer which is about 30 nm thick. A hard diamond-like carbon (DLC)

overcoat layer of thickness 1-3 nm exists on top of the magnetic layer to provide wear and

corrosion protection to the data. Above that is a thin layer (1-2 nm) of lubricant necessary

for wear protection and reliability. Figure 1.6 shows the head-gimbal assembly (HGA). The

slider has an air-bearing surface (ABS) with a specially designed etched surface (referred

to as ABS design). The suspension applies a fixed *preload* (or *gramload*) on the slider to

press it against the disk. When the disks spin, this preload balances the pressure generated

due to the air-bearing, causing the slider to fly over the disk surface.

The strength of the read-back signal depends on the magnetic spacing between

the head on the slider and the disk, also known as the head-media spacing (HMS). The

HMS includes the mechanical spacing, the slider and the disk overcoats and the protective lubricant layer on the disk. The mechanical spacing between the slider and the disk is also known as the flying height.

1.2   Evolution

The earliest *true* HDDs had the heads of the hard disk in contact with the surface of the disk. This was done to enable the low-sensitivity electronics of the day to read the magnetic fields on the surface of the disk. The key technological milestone that facilitated the creation of the modern hard disk drive came in the early 1950s. IBM engineers discovered that the heads could be made to float above the surface of the disk and read bits as they passed by underneath instead of rubbing continuously on the surface of the disk. The very first production hard disk drive was the IBM 350 disk drive system (sold with the IBM 305 RAMAC, Random Access Method of Accounting and Control, shown in Fig. 1.2, [Source: IBM corporation, 2008]), launched on September 13, 1956. This monster of a drive had 50 disks, each being 2 feet in diameter and could store 5 MB. Its areal density[1] was about 2,000 bits/in^2, along with a data transfer rate of 8,800 bytes/s.

Fast-forward to the year 2007. Hitachi GST announced the world's first 1 TB (Terabyte) drive. The Deskstar 7K1000 (see Fig. 1.3, [Source: Hitachi GST, 2008]) boasts an areal density of over 150 Gb/in^2 and data transfer rates of over 300 MB/s and costs less than $400. This clearly shows how over the past 50 years, the hard disk drive industry has made tremendous improvements in terms of storage density, performance and cost. The

---

[1] Areal recording density is defined as the amount of data that can be stored on a unit area of the recording media and is usually expressed as bits/in^2

evolution of hard disk drive areal density over the past 15 years can be seen from the chart

shown in Fig. 1.4 [Source: Seagate Technologies, 2008]. Starting with an areal density of 2

Kbit/in^2 (1 byte was 7 bits in those days) for the IBM 350 in 1956, the areal density has

grown exponentially with today's products exceeding 300 Gb/in^2. Note that the scale on

the $y - axis$ is logarithmic and not linear.

This exponential increase in hard disk drive capacity and access speeds has made

them viable for many consumer products and appliances that require large storage capaci-

ties, such as digital video recorders and digital audio players. An excellent summary of the

historical evolution of hard disk drives has been provided by Daniel, Mee, and Clark [1999].

1.3   Motivation

In the past decade, applications for HDDs have expanded far beyond computers

to varied and diverse applications such as digital video recorders, digital audio players,

personal digital assistants, digital cameras and video game consoles. As HDDs migrate

from desktop computers to such dynamic environments, the performance and reliability of

the HDD during adverse condition such as shocks, vibrations, extremes of temperature,

altitude and humidity become of great concern. The HDD that two decades back was

meant to run in a computer sitting stationary on a shelf for its lifetime, today has to sit in

ubiquitous digital audio players and keep up with the active lifestyles of their users. To aid

in the design process of HDDs, an ability to predict the response of the head-disk interface

during dynamic events such as load/unload (L/UL) and disturbances such as shock and

vibration becomes a big facilitator.

1.4 Outline

This dissertation is divided into 8 chapters. The first chapter gives an introduction of the hard disk drive, its temporal evolution as well as the future of the hard disk drive. It also includes a brief outline of the subsequent chapters, which is this section. The second chapter covers the modeling of the structural components of the hard disk drive, i.e. the suspension and the disk. In the third chapter we discuss modeling of contact at the head-disk interface, including asperity contacts and bulk interference contacts. A new boundary element method procedure is proposed to deal with bulk interference between the head and the disk. The fourth chapter deals with using the finite element method for the steady-state modeling of the air bearing. We also discuss methods for finding the steady-state flying attitude of the slider and propose a new method for doing the same. Strategies for grid refinement are investigated and convergence studies are also carried out. In chapter five, we extend the formulation for dynamic modeling of the air bearing. In chapter six, we use the framework developed in the previous four chapters to simulate shock events. Results from a previously developed simulator are presented investigating the effect of various parameters of shock on the performance of the suspension/air-bearing/disk system. In chapter seven, we use our framework to simulate the load/unload process as well as provide a comparative study with an earlier method for simulating the load/unload process. Finally in chapter eight, we outline conclusions we have drawn from this dissertation as well as chart out areas of further research.

1.5   Figures

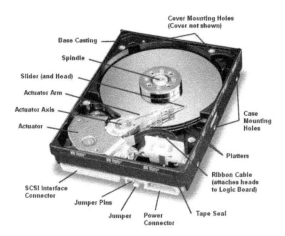

Figure 1.1: A typical modern hard disk drive

Figure 1.2: The IBM 350 Disk Storage System

Figure 1.3: The Hitachi Deskstar 7K1000

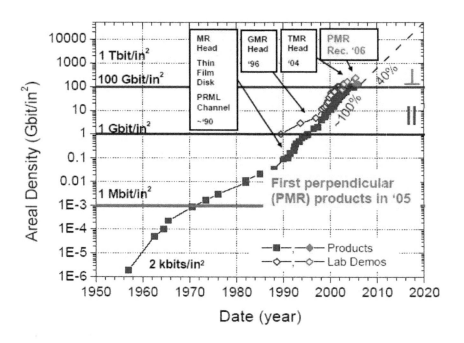

Figure 1.4: Evolution of areal density in hard disk drives

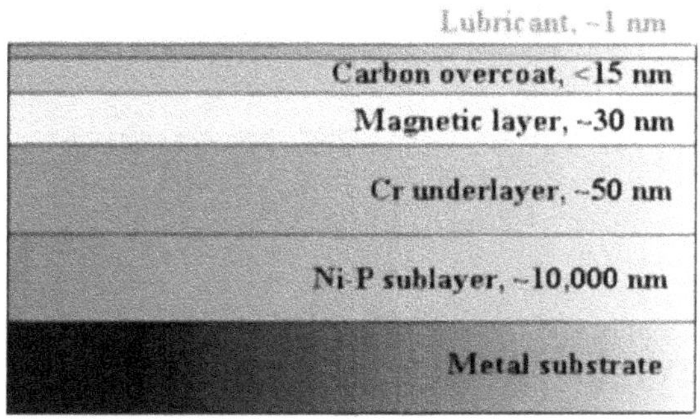

Lubricant, ~1 nm

Carbon overcoat, <15 nm

Magnetic layer, ~30 nm

Cr underlayer, ~50 nm

Ni-P sublayer, ~10,000 nm

Metal substrate

Figure 1.5: Various layers on the disk in a HDD

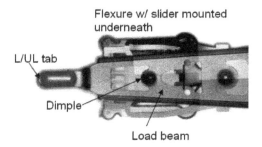

Flexure w/ slider mounted
underneath

L/UL tab

Dimple

Load beam

Figure 1.6: The head gimbal assembly

Chapter 2

Structural modeling

2.1 Introduction

In order to accurately simulate and predict the dynamic response of the head-disk interface of the hard disk drive, we require a model that not only captures the characteristics of the true system, but is also simple and computationally inexpensive to use. A good summary of previous works in this area has been provided by Gupta [2007]. Several researchers such as Ponnaganti [1986] and Hu and Bogy [1995] have used a simple HDI model to predict the system dynamics, which includes the slider model and an air bearing model which solves the generalized Reynolds equation. These models completely fail to accurately predict the dynamics of the HSA and the disk, due to the simple nature of the structural model used. According to Gupta [2007] the system dynamics predicted by these models is significantly different from the actual system response (measured experimentally) during slider-disk contact/impact, aerodynamic forcing on HSA due to disk rotation, shocks, track seek, load-unload etc. Gupta [2007] proposed a way to more accurately predict the system dynamics by including the HSA model along with the air bearing model that solves the generalized Reynolds equation.

In this chapter, we extend the model proposed by Gupta [2007] by including suspension nonlinearities such as flexure-dimple and limiter contacts. This extension enables us to simulate more complex dynamic processes in the hard disk drive such as load/unload, shock and vibration, where these nonlinearities become important. We also present a method for modeling the rotating disk using a similar procedure.

2.2   The HGA model

The Head gimbal assembly, or the HGA (which includes the suspension and the slider), is a structure that is used to load the slider (which houses the read/write heads) in close proximity to the rotating disk, such that the load applied by the suspension is balanced by the air bearing. A schematic of the suspension is shown in Fig. 2.2. A typical suspension consists of a stiff load beam and a flexure to which a slider is mounted. The load beam applies a load onto the flexure through the *dimple*. The dimple cannot support negative loads and is often observed to separate during events such as shock and load/unload. In order to prevent the dimple from opening up too much, there are limiters, which close when the dimple opens up a certain amount thereby preventing further separation between the load beam and the flexure (see Fig. 2.3).

2.3   Disk model

The disk is the second important structural component of the air-bearing system.
However in normal operation, the structural behavior of the disk is not of concern since
the deflections of the disk are very small. The effects of such small deflections due to disk
waviness and run-out can be included in the air-bearing calculations without modeling the
structural behavior of the disk itself.

The structural response of the disk becomes important during dynamic events such
as shock. However for the type of shocks we are modeling (*z-shocks*), the excitation of the
shock is symmetric about the rotation axis of the disk. It is well known that a rotating
disk has three types of modes. The first type are the axisymmetric modes with m nodal
circles and no nodal diameters $(m, 0)$, which have frequencies that only slightly increase
with the rotation speed, which is due to centrifugal stiffening. The first axisymmetric disk
mode is also referred to as the *umbrella* mode of vibration (Fig. 2.4). The frequencies
and mode shapes of the axisymmetric mode shapes are very similar in the operating and
non-operating states. The second are the asymmetric modes with zero nodal circles and n
nodal diameters $(0, n)$, each of which splits into two modes in the operating state (Fig. 2.5).
They have no response to symmetric excitations. The third are the coupled asymmetric
modes (m, n) where $(m > 0, n > 0)$ (Fig. 2.6), which are higher frequency modes, and their
responses are much smaller than the responses from the low frequency modes for ordinary
shocks (e.g., 0.5 ms half sinusoidal acceleration). A finite element model for the disk was
created in ANSYS using shell elements. The disk was subject to a 200 G shock for two cases:
when the disk was rotating at 3600 RPM and in the stationary state. Figure 2.7 shows the

z-displacement of a point on the outer periphery (OD) of the disk for both cases. Figure 2.8 shows the frequency spectra of the displacement. The first peak corresponds to the first axisymmetric mode of vibration of the disk. The other small peaks correspond to radial modes of vibrations of the disk, but the power contained within these is relatively very small. Thus we can see that since the disk response is predominantly axisymmetric, we can assume there is no significant difference in the shock responses of the disk during the operating and non-operating states when the shock is applied axially to the hub. In addition, the change in frequency due to the spinning of the disk is also negligible. Although the slider's load on the disk is not axisymmetric, it has been shown previously that its contribution to the disk response is negligible [Zeng and Bogy, 2000c]. Therefore, the disk responses to a symmetric shock are expected to be mainly from the axisymmetric modes, and they are similar in the operating and non-operating states. Hence, it is expected that this assumption will not cause a significant error in the simulation results. For non-axial shocks it will be important to consider the rotation of the disk while simulating the shock.

2.4 Governing Equation

In this section, we present the governing equation for the slider/air-bearing/disk system. When the slider is flying on the disk, it's motion is governed by the following equations:

$$\mathbf{M}_s \, \ddot{\mathbf{x}}_s + \mathbf{C}_s \, \dot{\mathbf{x}}_s + \mathbf{K}_s \, \mathbf{x}_s + \mathbf{F}_s = \mathbf{F}_{abs}(\mathbf{x}_s - \mathbf{x}_d, \dot{\mathbf{x}}_s - \dot{\mathbf{x}}_d) + \mathbf{F}_{con}(\mathbf{x}_s - \mathbf{x}_d) \qquad (2.1)$$

$$\mathbf{M}\,\ddot{\mathbf{x}} + \mathbf{C}\,\dot{\mathbf{x}} + \mathbf{K}\,\mathbf{x} + \mathbf{F}_{susp} = \mathbf{F}_{abs}(\mathbf{x}, \dot{\mathbf{x}}) + \mathbf{F}_{con}(\mathbf{x}) \qquad\qquad (2.2)$$

where $\mathbf{x}$ is the vector of the displacements of the degrees of freedom of the suspension, including the 6 degrees of freedom of the slider, namely displacements and rotations in the x, y and z directions. $\mathbf{M}$, $\mathbf{C}$ and $\mathbf{K}$ correspond to the mass, damping and stiffness matrices for the suspension. $\mathbf{F}_{susp}$ is the preload on the suspension, $\mathbf{F}_{abs}$ is the air-bearing force and $\mathbf{F}_{con}$ is the contact force between the slider and the disk. $\mathbf{F}_{abs}$ and $\mathbf{F}_{con}$ have non-zero values only at the degrees of freedom corresponding to those of the slider. These are calculated by integrating the air-pressure, shear stress, contact pressures and frictional stresses under the slider.

To calculate the stead state attitude of the slider, we neglect the time dependent terms in Eqn. 2.1 thus reducing it to:

$$\mathbf{K}\,\mathbf{x} + \mathbf{F}_{susp} = \mathbf{F}_{abs}(\mathbf{x}) + \mathbf{F}_{con}(\mathbf{x}) \qquad\qquad (2.3)$$

The above equation is solved directly to obtain the steady state attitude of the slider.

2.4.1   Linearization

Equations 2.1 and 2.3 are nonlinear in the displacement vector $\mathbf{x}$. In order to solve these equations, we linearize them and use the Newton Raphson scheme to solve them iteratively to obtain the solution to the corresponding nonlinear problems. Linearizing the dynamic governing equation Eqn. 2.1 with respect to $\mathbf{x}$ about a point $\mathbf{x_0}$, we obtain:

$$\mathbf{M}\,\ddot{\partial \mathbf{x}} + \mathbf{C}\,\dot{\partial \mathbf{x}} + \mathbf{K}\,\partial \mathbf{x} - \mathbf{C}_{abs}\,\dot{\partial \mathbf{x}} - \mathbf{K}_{abs}\,\partial \mathbf{x} - \mathbf{K}_{con}\,\partial \mathbf{x} =$$

$$- \mathbf{M}\,\ddot{\mathbf{x}}_0 - \mathbf{C}\,\dot{\mathbf{x}}_0 - \mathbf{K}\,\mathbf{x_0} - \mathbf{F}_{susp} + \mathbf{F}_{abs}(\mathbf{x_0}, \dot{\mathbf{x}}_0) + \mathbf{F}_{con}(\mathbf{x_0}) \quad (2.4)$$

where $\partial \mathbf{x} = \mathbf{x} - \mathbf{x_0}$, $\mathbf{C}_{abs}$ and $\mathbf{K}_{abs}$ are the damping and stiffness matrices associated with the air bearing and $\mathbf{K}_{con}$ is the stiffness matrix associated with contact. Similarly the linearized form of the steady state equation Eqn. 2.3 works out to:

$$\mathbf{K}\,\partial \mathbf{x} - \mathbf{K}_{abs}\,\partial \mathbf{x} - \mathbf{K}_{con}\,\partial \mathbf{x} = -\mathbf{F}_{susp} + \mathbf{F}_{abs}(\mathbf{x_0}) + \mathbf{F}_{con}(\mathbf{x_0}) \quad (2.5)$$

Solving this equation iteratively yields the solution to the nonlinear steady state inverse problem.

2.5   Guyan reduction

HSA models used in today's drives are highly complex high performance structures, designed to satisfy stringent dynamic performance requirements such as shock. Typical finite element models of the HSA (see Fig. 2.1) can include over 100,000 nodes, which results in a huge number of degrees of freedom. Solving such a large system coupled nonlinearly with the fluid dynamic air bearing makes our problem of evaluating the dynamic performance of the air bearing slider very expensive.

In order to make our problem more tractable, we use the technique of Dynamic reduction to reduce the total number of degrees of freedom of our suspension system. Perhaps the most widely used dynamic reduction technique is Guyan reduction [Guyan, 1965].

A good survey of newer, more accurate dynamic reduction techniques has been presented

by Gupta [2007]. However for our problem, the original method proposed by Guyan [1965]

is sufficiently accurate and is used here. Moreover we are using ANSYS, which is a com-

mercially available finite element modeling program to model our suspension, and Guyan

reduction is already available in ANSYS in the form of *substructuring*. We use ANSYS

to generate mass and stiffness matrices for our dynamic computations [see Bhargava and

Bogy, 2006].

2.6   Modeling suspension nonlinearities

As discussed in section 2.2, even though the HSA is a linear elastic system, its

response is nonlinear due to contacts at the dimple and limiter. After exporting the linear

structural mass and stiffness matrices from ANSYS with no contacts, we impose the contact

conditions on the matrices by using a Lagrange multiplier method. However simply imposing

the contact and no-contact conditions is not sufficient, we also require impact and release

conditions for the contacting nodes. We use the contact and impact conditions proposed by

Hughes, Taylor, Sackman, Curnier, and Kanoknukulchai [1976]. In that paper the authors

derive discrete impact and release conditions for structural finite-element models with node

to node contacts using the results from the impact of two elastic rods.

2.7 Time discretization

For discretizing the equation in time, we use the Newmark-Beta method [see Paz and Leigh, 2004]. Time-step size selection is accomplished by employing an a posteriori error estimator proposed by Zienkiewicz and Xie [1991]. Time-step size is also reduced during status change of contact elements and when the slider crashes into the disk.

2.8 Discussion

The modeling of the structural components of the hard disk drive system were discussed in this chapter. We propose to use a stationary disk with centrifugal stiffening to model the axisymmetric response of a rotating disk. For the suspension model, we propose to use pre-assembled reduced mass and stiffness matrices from ANSYS, with nonlinear contact constraints imposed on top of these matrices using a Lagrange multiplier method and impact/release conditions.

2.9 Figures

Figure 2.1: An HSA model for a popular 1" drive

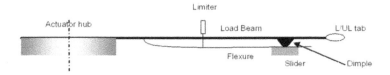

Figure 2.2: Schematic of the HSA

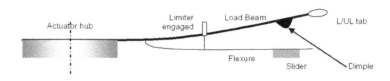

Figure 2.3: Schematic of the HSA showing limiters engaged

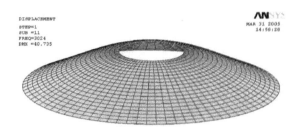

Figure 2.4: First axisymmetric (umbrella) mode of the disk

Figure 2.5: First radial mode of the disk

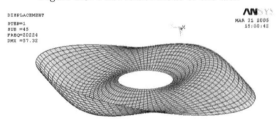

Figure 2.6: First axisymmetric-radial coupled mode

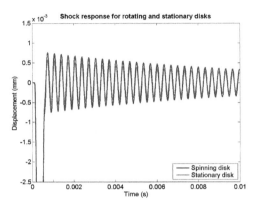

Figure 2.7: Disk response to z-shock for spinning and stationary disks

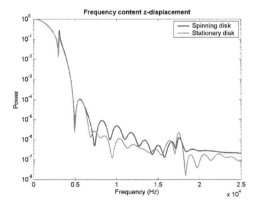

Figure 2.8: Power spectra of disk shock response for spinning and stationary disks

Chapter 3

Contact modeling

3.1 Introduction

In hard disk drives, failure at the head disk interface is often caused by head-disk contacts leading to permanent damage of either the head or the disk or both as well as loss of data. The seriousness of a head disk-impact during events such as shock and load/unload can be characterized by the magnitude of the contact forces and pressures observed during the course of the contact, thus accurately modeling the contact phenomenon during such events is of utmost importance.

We define the net contact force between the head and the disk as a sum of two forces, the *asperity contact* force and the *impact* force. The asperity contact force is the force due to the roughness of the contacting surfaces, i.e. the force due to the contacting asperities on the two surfaces. The impact force on the other hand is the force generated due to bulk interference between the two contacting surfaces.

3.2 Asperity Contact forces

The asperity contact forces are calculated using the Greenwood-Williamson statistical asperity contact model [see Greenwood and Williamson, 1966].

3.3 Impact forces

For modeling the impact forces, we develop a new boundary element method based quasi-static impact model. The details of this model are presented in this section.

3.3.1 Assumptions

The problem of two bodies in contact is nonlinear due to the nonlinear boundary conditions of contact. Considering the problem of the slider impacting the disk, we make the following assumptions:

1. Small deformations: The deformations due to the slider-disk interactions are assumed to be small such that linear elasticity theory is applicable.

2. Quasi-static behavior: We assume the behavior of the slider and the disk is close to quasi-static contact, such that the slider and disk are in equilibrium at each instant of the impact. This assumption is valid for most contacts, since the speed of impact ($\sim$ 1m/s) is much lower than the speed of sound in the materials. This assumption allows us to independently calculate contact pressures at each time step given the slider attitude.

3. Rigid slider: Since the slider (usually made of AlTiC, $E \approx 390\text{MPa}$) is much stiffer than the substrate of the disk (Aluminum or glass, $E \approx 70\text{MPa}$), we can assume the deformation of the slider is negligible as compared to the disk. The effect of slider deformation can however be partially included by defining an effective Young's modulus for the disk, which takes into account the compliance of the slider.

4. Effect of overcoats is negligible: The effects of various overcoats on the disk such as the magnetic layer, TLC, etc., is neglected due to their small thickness ($\sim 25\text{nm}$).

5. Half space assumption: The disk is modeled as a half-space because the length scale for the thickness of the disk is 4 to 5 orders of magnitude larger than the scale of the deformation.

3.3.2 Formulation

Under the assumptions listed above, the governing equation for the slider disk system becomes:

$$\mathbf{M\ddot{x}} + \mathbf{C\dot{x}} + \mathbf{Kx} = \mathbf{F_{abs}}(\mathbf{x}, \dot{\mathbf{x}}) + \mathbf{F_{con}}(\mathbf{x}) + \mathbf{F_{imp}}(\mathbf{x}) \tag{3.1}$$

where $\mathbf{M}$, $\mathbf{C}$ and $\mathbf{K}$ refer to the mass, damping and stiffness for the slider suspension system, and $\mathbf{F_{abs}}$, $\mathbf{F_{con}}$ and $\mathbf{F_{imp}}$ are the air bearing, asperity contact and impact forces. The above equation is linearized about $(\mathbf{x_0}, \dot{\mathbf{x}}_0, \ddot{\mathbf{x}}_0)$ to get:

$$\mathbf{M\ddot{x}} + \mathbf{C\dot{x}} + \mathbf{Kx} = \mathbf{F_{abs}}(\mathbf{x_0}, \dot{\mathbf{x}}_0) + \mathbf{K_{abs}}\partial\mathbf{x} + \mathbf{C_{abs}}\partial\dot{\mathbf{x}}$$
$$+ \mathbf{F_{con}}(\mathbf{x_0}) + \mathbf{K_{con}}\partial\mathbf{x} + \mathbf{F_{imp}}(\mathbf{x_0}) + \mathbf{K_{imp}}\partial\mathbf{x} \tag{3.2}$$

where $\partial \mathbf{x} = \mathbf{x} - \mathbf{x_0}$. Thus for the impact force calculations, we need to find the impact force at any given attitude $\mathbf{F}_{imp}(\mathbf{x})$ and also to find the impact stiffness at an given attitude $\mathbf{K}_{imp}(\mathbf{x})$.

3.3.3   The Boussinesq problem

The Boussinesq problem is that of a concentrated force $\mathbf{F}$ applied on an unbounded elastic medium (see Fig. 3.1). The exact solution for the particular Boussinesq problem [see Lur'e, 2005] wherein the concentrated force is applied normal to the half space is used in our formulation. The displacement in the half space due to the concentrated force of magnitude F at the origin are given by:

$$\mathbf{u} = \frac{F}{4\pi\mu}\left\{\frac{1}{R}\left[(3-4\nu)\mathbf{i_3} + \frac{z}{R^3}\mathbf{R}\right] - (1-2\nu)\nabla\ln(R+z)\right\} \qquad (3.3)$$

Simplifying and evaluating the z-displacement at the surface (i.e. $z = 0$), we get:

$$d = \frac{F(1-\nu)}{2\mu\pi R} \qquad (3.4)$$

We use Eqn. 3.4 as the Green's function in our BEM formulation.

3.3.4   The Boundary Element Method

Consider the case when the slider (denoted by S) and the disk (assumed to be a half space, denoted by D) are in contact, as shown in Fig. 3.2. Let the contact region be denoted by C, such that $C = \partial A \cap \partial B$. Our problem is to determine the contact pressure $\mathbf{p}$ over the region C, given the attitude of the slider (and hence the deformation of the disk d). The region C is not known initially, so we assume that the area of the slider below the undeformed surface of the disk (i.e. the region on the slider with negative fly-height) is in contact with the disk. To apply the boundary element method, we discretize the contact pressures as having the form:

$$\mathbf{p} = \sum_{i=1}^{N_n} p_i \phi_i \tag{3.5}$$

where p_i are the unknowns and ϕ_i form our set of basis functions. We also assume the same form for the deformations:

$$d = \sum_{i=1}^{N_n} d_i \phi_i \tag{3.6}$$

For our solution, we discretize (mesh) our contact domain C into triangular elements T_i such that $C = \bigcup T_i$. We use piecewise linear functions over these elements as our basis functions (see Fig. 3.3). This is consistent with the finite element discretization for the spacing and air bearing pressures for the generalized Reynold's equation, in case such a method is employed. Now from the solution to the Boussinesq problem, we know that the deformation at a location (x_p, y_p) due to a pressure p at a location (x, y) over an area $\mathrm{d}A$ can be calculated from Eqn. 3.3 as:

$$d(x_p, y_p) = \frac{p(1 - \nu)\,\mathrm{d}A}{2\mu\pi\sqrt{(x_p - x)^2 + (y_p - y)^2}} \tag{3.7}$$

Thus the deformation at (x_p, y_p) due to the complete contact pressure distribution over the region C is:

$$\begin{aligned} d(x_p, y_p) &= \int_C \frac{p(1 - \nu)\,\mathrm{d}A}{2\mu\pi\sqrt{(x_p - x)^2 + (y_p - y)^2}} \\ &= \sum_{j=1}^{N_e} \int_{T_j} \frac{p(1 - \nu)\,\mathrm{d}A}{2\mu\pi\sqrt{(x_p - x)^2 + (y_p - y)^2}} \end{aligned} \tag{3.8}$$

For each element, we can map the physical element onto a reference element with barycentric coordinates ψ and η (see Fig. 3.4). Substituting the form of our contact pressure assumed in Eqn. 3.5 into Eqn. 3.8, it becomes:

$$d(x_p, y_p) =$$

$$\sum_{j=1}^{N_e} \left[\frac{\triangle_j}{\mu \pi} \int_0^1 \int_0^{1-\psi} \frac{[p_1(1-\eta-\psi) + p_2\eta + p_3\psi](1-\nu) \, \mathrm{d}\eta \, \mathrm{d}\psi}{\sqrt{(x_1(1-\eta-\psi) + x_2\eta + x_3\psi - x_p)^2 + (y_1(1-\eta-\psi) + y_2\eta + y_3\psi - y_p)^2}} \right]$$

$$(3.9)$$

Although the above integral can be evaluated numerically, a high order quadrature

is required for an accurate value, especially when deformations are being evaluated at a

point which lies close to the element. This would be numerically very expensive. Hence the

integral is evaluated analytically using Mathematica. Evaluating the integral, we get:

$$d_p = \sum_{i=1}^{N_n} c_{pi} \, p_i \qquad (3.10)$$

where c_{pi} represents the deformation at node p due to the pressure at node i. Assembling

the equations at each node, we get a system of equations in the unknown nodal pressures

p_i with the nodal deformations d_i on the right hand side as:

$$\{\mathbf{d}\} = [\mathbf{K}] \, \{\mathbf{p}\} \qquad (3.11)$$

$$\Rightarrow \{\mathbf{p}\} = [\mathbf{K}]^{-1} \, \{\mathbf{d}\} \qquad (3.12)$$

The size of this system of equations depends on the number of nodes interfering

with the disk, which depends on the severity of the contact. This system of equations can

be solved by Gaussian elimination (LU) to obtain the unknown nodal contact pressures.

Initially we had assumed that the entire region of the slider with negative fly-height

is in contact with the disk. However in reality this will not be the case. After solving the

set of equations in Eqn. 3.11, we obtain negative values of pressure at some of the nodes.

These are the points which are not in contact with the disk. Thus these points must be removed from the system of equations and the remaining equations must be resolved. This procedure is a sort of fixed point iteration for the nonlinear contact problem.

After solving for the impact pressures, we can integrate the pressures over all the elements to obtain contact forces and moments using:

$$F_z = \sum_{j=1}^{N_e} \int_{T_j} p(x,y) \, dA \qquad (3.13)$$

$$M_{pitch} = \sum_{j=1}^{N_e} \int_{T_j} p(x,y) \cdot x \, dA \qquad (3.14)$$

$$M_{roll} = \sum_{j=1}^{N_e} \int_{T_j} p(x,y) \cdot y \, dA \qquad (3.15)$$

3.3.5 Calculation of impact stiffness

As seen in our governing equations in Eqn. 3.2, we also need to find the impact stiffness $\mathbf{K_{imp}}$. The stiffness here is the variation in contact forces and moments due to variations in the attitude of the slider (i.e. the z-height, pitch and the roll). We can write $\mathbf{K_{imp}}$ as being composed of two components: one due to the change in contact pressure over the existing contact region and the second due to the change in the contact region itself. Since the slider has sharp etched features, we assume that the component of the stiffness due to the change in contact area is negligible, i.e. the contact area does not change for small variations in slider attitude. Thus the contact stiffness is simply the change in contact pressures due to variations in slider attitude. Let the parameter α_i refer to the z-height, pitch and the roll for $i = 1, 2, 3$. We need to evaluate $\frac{\partial p_j}{\partial \alpha_i}$. We take Eqn. 3.11 and differentiate both sides with respect to α_i:

$$\left\{ \frac{\partial \mathbf{p}}{\partial \alpha_i} \right\} = [\mathbf{K}]^{-1} \left\{ \frac{\partial \mathbf{d}}{\partial \alpha_i} \right\} \tag{3.16}$$

Our system matrix $\mathbf{K}$ is independent of the slider attitude because we are neglecting the change in the contact region due to small variations in the slider attitude. Looking at the definition of d_i:

$$d_i = h_{0_i} + \alpha_1 + x \cdot \alpha_2 + y \cdot \alpha_3 \tag{3.17}$$

for small values of pitch (α_2) and roll (α_3). Thus the derivatives with respect to the slider attitude become:

$$\frac{\partial d_i}{\partial \alpha_1} = 1 \tag{3.18}$$

$$\frac{\partial d_i}{\partial \alpha_2} = x \tag{3.19}$$

$$\frac{\partial d_i}{\partial \alpha_3} = y \tag{3.20}$$

Thus by solving three additional linear systems (which is cheap if we're using a direct method, since $\mathbf{K}$ has already been factorized), we can obtain the pressure variation due to the slider's attitude. Now to calculate the changes in contact forces and moments, we simply substitute $\frac{\partial p}{\partial \alpha_i}$ for p in Eqns. 3.13, 3.14 and 3.15. Thus we get:

$$\frac{\partial F_z}{\partial \alpha_i} = \sum_{j=1}^{N_e} \int_{T_j} \frac{\partial p}{\partial \alpha_i} \, dA \tag{3.21}$$

$$\frac{\partial M_{pitch}}{\partial \alpha_i} = \sum_{j=1}^{N_e} \int_{T_j} \frac{\partial p}{\partial \alpha_i} \cdot x \, dA \tag{3.22}$$

$$\frac{\partial M_{roll}}{\partial \alpha_i} = \sum_{j=1}^{N_e} \int_{T_j} \frac{\partial p}{\partial \alpha_i} \cdot y \, dA \tag{3.23}$$

3.3.6 Radius of effect

The **K** in the system of equations obtained from the above described procedure (Eqn. 3.11) is full, typically with no zero terms. This is because contact pressure at one location will cause deformation at all other locations on the contact region. This makes the process of forming the equations very expensive. To reduce this computation, we can make the assumption that the contact pressure at one location will only cause a deformation within a radius r_e around it, which we refer to as the 'radius of effect'. This simplification significantly cuts down the computation time for the contact pressures. The effect of this parameter r_e (normalized with respect to indenter size, a_0) on the contact force is plotted in Fig. 3.5 for a flat-end circular indenter indenting on a half plane. We find that the error in the net contact force is about 35% for $r_e = 0.5a_0$ and reduces to less than 12% for $r_e = 0.8a_0$. This simplification also makes the matrix **K** sparse. This sparsity can also be exploited during the solution of the system of equations.

3.3.7 Solution comparison

In this section, we compare the results of the new method with exact analytical solutions for problems of an axisymmetric punch [Fu and Chandra, 2002]. Meshes for the boundary element method are generated using Triangle [Shewchuk, 1996].

Flat-end circular indenter

The pressure profile for a flat-ended circular indenter is given by:

$$p(r) = -\frac{a_0}{\pi} \cdot \frac{E}{1 - \nu^2} \cdot \frac{1}{\sqrt{a^2 - r^2}} \tag{3.24}$$

where a_0 is the penetration depth and a is the radius of the indenter. The exact and BEM

impact pressure profiles are plotted in Fig. 3.6. We observe that the two solutions are in

good agreement.

Parabolic indenter

Consider a parabolic indenter where the vertical displacement profile is defined as:

$$d(r) = a_0 + a_2 r^2 \ \ (0 \leq r \leq a) \tag{3.25}$$

where a is again the radius of the indenter. Figure 3.7 shows a plot of the radial pressure

profile under a parabolic indenter for various penetration depths. The middle set of curves

shows critical complete contact, wherein there is no singularity at the indenter edge and

for the top set of curves, penetration depth is larger than the critical depth and there is a

square root singularity at the indenter edge in the analytical solution. In Fig. 3.8, we plot

the total force under the indenter. We observe that the BEM based model agrees well with

the analytical solution.

Hard disk air bearing slider

Figure 3.9 shows the geometry of the air bearing surface of a hard disk slider. The

CML Finite Element Dynamic Air Bearing simulator is a computer program used to solve

the air bearing equations for such a slider [Bhargava and Bogy, 2007a]. Figure 3.10 shows a

mesh used for the solution of the generalized Reynolds equation over the slider. We use the

same mesh in the program for the solution of contact pressures. Figures 3.11-3.13 plot the impact pressures for varying slider attitudes. We see that the peak pressures are observed at the edges of the interfering rails and not in areas of maximum penetration (as expected). The forces and moments calculated by integrating the pressures are included in the dynamic simulations to calculate the accurate response of the slider during head-disk contacts.

3.4 Discussion

A new boundary element method based quasi-static impact model is developed for slider-disk impacts in hard disk drives. The model is used to calculate impact forces as well as impact stiffness, and it is used in the CML FE Dynamic Air Bearing simulator to accurately model slider-disk impacts. The impact pressures calculated are compared with analytical solutions and are found to be in good agreement.

3.5   Figures

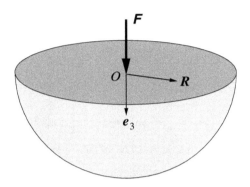

Figure 3.1: The Boussinesq problem

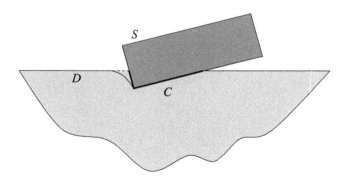

Figure 3.2: Slider disk contact

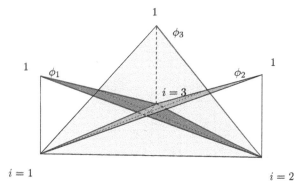

Figure 3.3: Elemental basis functions

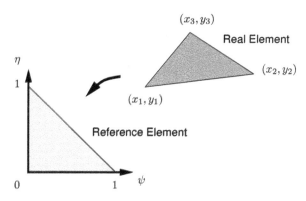

Figure 3.4: Reference element

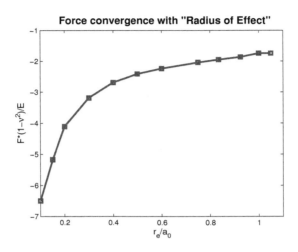

Figure 3.5: Convergence as a function of 'Radius of Effect'

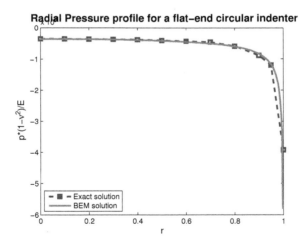

Figure 3.6: Radial pressure profile for a rigid flat-end circular indenter

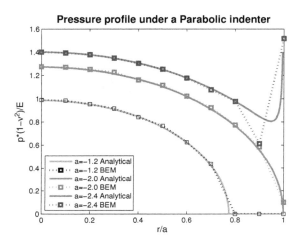

Figure 3.7: Radial pressure profile for a rigid parabolic circular indenter

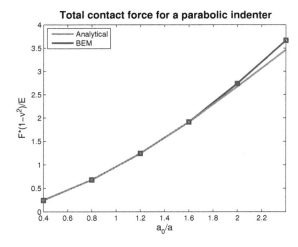

Figure 3.8: Total contact force for a rigid parabolic circular indenter

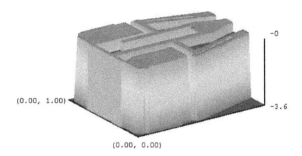

Figure 3.9: Slider profile

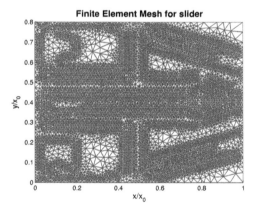

Figure 3.10: Finite element mesh

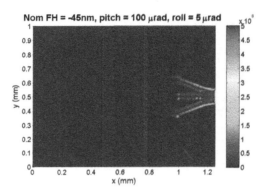

Figure 3.11: Pressure profile during Slider-disk contact 1

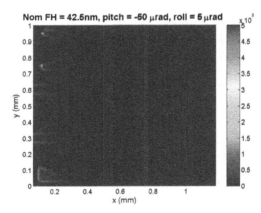

Figure 3.12: Pressure profile during Slider-disk contact 2

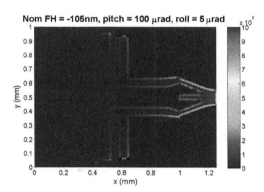

Figure 3.13: Pressure profile during Slider-disk contact 3

Chapter 4

Fluid static modeling

4.1 Introduction

The hard disk drive industry continually faces a demand for higher data areal densities and access speeds. This in turn translates into a need for lower fly heights for the read/write heads. The lower fly heights and the demand for increased reliability forces the design of extremely complex slider geometries with steep etches that define the air bearing surface shapes. In order to evaluate such designs, there is a need for fast, reliable simulations of their flying characteristics. Thus there is a need for a simulation program which can solve for the flying attitudes of complicated slider designs as quickly as possible.

Before considering the transient solution for the air bearing, we first look at the behavior of the air bearing slider in the steady state for two reasons:

1. The primary criterion used to evaluate the design of any air bearing slider are the attitude of the slider in the steady state and the stiffness of the air bearing.

2. The solution to the steady state problem is often used as an initial condition to many dynamic problems such as shock and slider unloading. Hence the solution to the steady state problem is important. The alternative of solving a transient analysis until steady state is reached is a very expensive proposition.

4.2   Previous Work

The problem of obtaining the static fly height of hard disk drive air-bearing sliders has previously been solved by various researchers in academia and industry. An excellent survey of these works was presented by Holani and Müftü [2005]. These studies have used the finite difference [White and Nigam, 1980; Castelli and Pirvics, 1968], finite volume [Hu and Bogy, 1995; Wu and Bogy, 2000; Cha and Bogy, 1995] and finite element [Smith, Wahl, and Talke, 1995; Garcia-Suarez, Bogy, and Talke, 1984; Hendricks, 1988; Kubo, Ohtsubo, Kawashima, and Marumo, 1988; Peng and Hardie, 1995] methods for solving this problem. The equilibrium attitude of the slider depends on the suspension forces acting on the slider and the air pressure due to the air-bearing. Various methods have been proposed to accomplish the coupling between the two sets of equations (structural and air-bearing). One such approach includes the dynamic effects of the air bearing (i.e. the squeeze term for the Reynolds equation and inertial effects for the slider) [Tang, 1972; Ono, 1972; White and Nigam, 1980; Miu and Bogy, 1986; Ruiz and Bogy, 1990; Cha and Bogy, 1995]. The steady state flying attitude is found by computing the transient solution until a steady state condition has been achieved. Another completely different approach obtains the coupled solution by formulating the problem for the steady state by neglecting the time-dependent terms [Yamaura and Ono, 1990; Choi and Yoon, 1994; Smith et al., 1995]. As stated by Holani and Müftü [2005], probably the most challenging problem with numerical methods is the need to discretize a continuous domain into one with finite discrete parameters. The finite difference method typically requires a structured mesh and is thus limited in its mesh refinement capabilities. Wu and Bogy [2000] presented a finite volume method with unstructured triangular meshing to solve the Generalized Reynolds Equation. A finite element method does not require a structured mesh. This offers a major advantage in that it renders the method highly suitable for local adaptive mesh refinement using geometry and pressure gradients.

In this paper we present a new formulation for dealing with the *inverse problem* wherein the nonlinear structural-fluid coupling is linearized along with the Reynolds equation to solve just one set of iterations, instead of two (solving for the slider attitude in one, and solving the nonlinear Reynolds equation in the other). We also present a new method of obtaining air bearing stiffness wherein the exact stiffness matrix is obtained by solving a set of 3 linear problems, as opposed to the conventional perturbation method where three nonlinear systems are required to solve for the directional stiffness and 6 nonlinear systems are needed to solve for centered stiffness matrices. We also present a new hybrid strategy for mesh refinement which uses pressure gradient and pressure gradient discontinuities across elements to give rapid spatial grid convergence.

4.3 Methodology

4.3.1 Governing equations

When the slider is flying on the disk, it's motion if governed by the following equation:

$$\mathbf{M}\,\ddot{\mathbf{x}} + \mathbf{C}\,\dot{\mathbf{x}} + \mathbf{K}\,\mathbf{x} + \mathbf{F}_{susp} = \mathbf{F}_{abs}(\mathbf{x}, \dot{\mathbf{x}}) + \mathbf{F}_{con}(\mathbf{x}) \tag{4.1}$$

where x is the vector of the 6 degrees of freedom of the slider, namely displacements and rotations in the x, y and z directions. To calculate the steady state (static) flying attitude, we neglect the time dependent terms, to get:

$$\mathbf{K}\,\mathbf{x} + \mathbf{F}_{susp} = \mathbf{F}_{abs}(\mathbf{x}) + \mathbf{F}_{con}(\mathbf{x}) \tag{4.2}$$

The air-bearing force is calculated by integrating the air-pressure and the shear stress under the slider. The Generalized Reynolds Equation is solved to get the pressure field. However the spacing between the slider and the disk is extremely small (of the order of a few nm, which is much less than the mean free path of air). Under these conditions, the continuum assumption for the air and the no-slip boundary conditions are no longer valid. Hence the Reynolds equation is modified to include rarefaction and slip effects. The modified Generalized Reynolds Equation can be written in the following form in terms of dimensionless variables, $P = p/p_0$ (pressure normalized with respect to the ambient pressure p_0), $H = h/h_m$ (slider-disk clearance normalized with respect to a nominal spacing h_m) and $T = \omega t$ (time non-dimensionalized with respect to the angular velocity of the disk ω):

$$\nabla \cdot (Q\,PH^3\,\nabla P) = \mathbf{\Lambda} \cdot \nabla(PH) + \tau\frac{\partial}{\partial T}(PH) \tag{4.3}$$

where $\nabla = \frac{\partial}{\partial X}\mathbf{E}_X + \frac{\partial}{\partial Y}\mathbf{E}_Y$ is the gradient operator with respect to the normalized coordinates $X = x/L$ and $Y = y/L$ where L is the characteristic length scale for the slider. The non-dimensional vector $\mathbf{\Lambda}$ is the bearing number defined as $\mathbf{\Lambda} = \frac{6\mu\mathbf{U}L}{p_0 h_m^2}$, where μ is the dynamic viscosity of air and $\mathbf{U}$ is the velocity of the disk. The squeeze number τ is defined as $\tau = \frac{12\mu\omega L^2}{p_0 h_m^2}$. It represents the ratio of transient effects to the diffusion effects in the problem. The factor Q is the modification to the continuum Generalized Reynolds Equation for incorporating slip and rarefaction effects. Various formulae have been proposed for Q to implement rarefaction and slip effects. A good discussion of the various models has been given by Wu [2001]. Values of the parameter Q for different models are briefly listed in Table 4.1.

Table 4.1: Rarefaction and slip models

Model	Flow Factor Q
Continuum model	1
First order slip model	$1 + 6a\frac{K_n}{PH}$
Second order slip model	$1 + 6\frac{K_n}{PH} + 6\left(\frac{K_n}{PH}\right)^2$
Fukui-Kaneko model	$f\left(\frac{K_n}{PH}\right)$

In the table above, $a = \frac{2-\alpha}{\alpha}$, where α is the accommodation factor and $K_n = \frac{\lambda}{h_m}$ is the Knudsen number, λ being the mean free path of air. The function $f\left(\frac{K_n}{PH}\right)$ is given by Fukui and Kaneko [1990]. To calculate the steady state pressure distribution over the slider, we neglect the time dependent terms to get the steady state generalized Reynolds equation:

$$\nabla \cdot (Q\, PH^3\, \nabla P) = \mathbf{\Lambda} \cdot \nabla(PH) \tag{4.4}$$

The above equation is solved over the domain of the slider (denoted by $\mathcal{S}$) with the boundary conditions of ambient pressure $(P = 1)$ at the edges of the slider to get the pressure distribution over the slider.

4.3.2 The Weak Form

To derive the weak form of Eqn. 4.4, we multiply the equation by a test function v, integrate over the domain $\mathcal{S}$ and use the divergence theorem to get:

$$\int_{\mathcal{S}} \nabla v \cdot (Q \, PH^3 \, \nabla P) \, \mathrm{dA} + \int_{\mathcal{S}} v \, \mathbf{\Lambda} \cdot \nabla(PH) \, \mathrm{dA} - \int_{\partial\mathcal{S}} v \, (Q \, PH^3 \, \nabla P) \cdot \mathbf{n} \, \mathrm{dS} = 0 \quad (4.5)$$

The above equation along with the boundary condition $P = 1$ over $\partial\mathcal{S}$ is the weak form of Eqn. 4.4 and can be solved to give the pressure field over the slider $\mathcal{S}$. We decompose the domain of the slider $\mathcal{S}$ into a finite number of triangular domains called 'elements', such that:

$$\mathcal{S} = \bigcup_{i=1}^{N_e} \mathcal{T}_i \;\; || \; \mathcal{T}_i \cap \mathcal{T}_j = \emptyset \text{ for } i \neq j \quad (4.6)$$

Writing the weak form Eqn. 4.5 over each element $\mathcal{T}_i$, we have:

$$\int_{\mathcal{T}_i} \nabla v \cdot (Q \, PH^3 \, \nabla P) \, \mathrm{dA} + \int_{\mathcal{T}_i} v \, \mathbf{\Lambda} \cdot \nabla(PH) \, \mathrm{dA} - \int_{\partial\mathcal{T}_i} v \, (Q \, PH^3 \, \nabla P) \cdot \mathbf{n} \, \mathrm{dS} = 0 \quad (4.7)$$

Now since pressure boundary conditions are being applied to the boundary of the slider (i.e. pressure is known on $\partial\mathcal{S}$), we set $v = 0$ over $\partial\mathcal{S}$. Thus we have:

$$\int_{\mathcal{T}_i} \nabla v \cdot (Q \, PH^3 \, \nabla P) \, \mathrm{dA} + \int_{\mathcal{T}_i} v \, \mathbf{\Lambda} \cdot \nabla(PH) \, \mathrm{dA} - \int_{\partial\mathcal{T}_i \backslash \partial\mathcal{S}} v \, (Q \, PH^3 \, \nabla P) \cdot \mathbf{n} \, \mathrm{dS} = 0 \quad (4.8)$$

This is the elemental weak form of the equation.

4.3.3 The Forward Problem

In the forward problem the objective is to find the pressure distribution given the attitude of the slider. Since the attitude is specified and the slider profile is known, H over the slider S is known, thus the only unknown in the problem is P over the slider S.

Linearization

To solve the nonlinear forward problem by the Newton Raphson method, Eqn. 4.8 needs to be linearized with respect to the pressure P. Writing $P = P_0 + \partial P$ and retaining only the linear terms in ∂P we get:

$$\int_{T_i} \nabla v \cdot (Q \, P_0 H^3 \, \nabla P_0 + Q \, P_0 H^3 \, \nabla \partial P + Q \, \partial P H^3 \, \nabla P_0) \, \mathrm{dA}$$

$$+ \int_{T_i} v \, \boldsymbol{\Lambda} \cdot (H \, \nabla P_0 + P_0 \, \nabla H + H \, \nabla \partial P + \partial P \, \nabla H) \, \mathrm{dA}$$

$$- \int_{\partial T_i \backslash \partial S} v \, (Q \, P_0 H^3 \, \nabla P_0 + Q \, P_0 H^3 \, \nabla \partial P + Q \, \partial P H^3 \, \nabla P_0) \cdot \mathbf{n} \, \mathrm{dS} = 0 \quad (4.9)$$

The above equation gives the linearized form of the elemental weak form. The linearized form will later be formulated in the form of Newton's iterations to obtain the solution to the nonlinear problem.

Finite element discretization

In order to solve the pressure field problem numerically, we approximate it by assuming the form:

$$P = \sum_{i=1}^{N_n} P_i \phi_i \qquad (4.10)$$

where P_i are the unknowns and ϕ_i are basis functions. We use piecewise linear functions

over triangular elements as our basis functions (see Fig. 4.3) where the P_i correspond to

the values of the pressure at the vertices of the elements (i.e. the nodes). Locally over each

element, for our bilinear basis functions, we have $P = P_i \phi_i^e$ for $i = 1, 2, 3$, where ϕ_i^e are

the local basis functions corresponding to node i. We denote the nodal pressures and basis

functions for each element by the vectors $\mathbf{P}^e = P_i$ and $\boldsymbol{\phi}^e = \phi_i$. Thus $P = \boldsymbol{\phi}^{eT} \mathbf{P}^e$. The

pressure gradient becomes $\nabla P = \mathbf{B}^{eT} \mathbf{P}^e$ where $\mathbf{B}^e = \left\{ \frac{\partial \phi_i^e}{\partial X} \quad \frac{\partial \phi_i^e}{\partial Y} \right\}$.

For the test functions, we use a different set of basis functions $\tilde{\phi}_i$, such that:

$$v = \sum_{i=1}^{N_n} v_i \tilde{\phi}_i \tag{4.11}$$

Locally over each element, again we have $v = v_i \tilde{\phi}_i^e$ for $i = 1, 2, 3$ where $\tilde{\phi}_i^e$ are the

local test basis functions corresponding to node i. We denote the nodal test function value

and basis functions for each element by the vectors $\mathbf{v}^e = v_i$ and $\tilde{\boldsymbol{\phi}}^e = \tilde{\phi}_i$. Thus $v = \tilde{\boldsymbol{\phi}}^{eT} \mathbf{v}^e$.

The gradient of the test function becomes $\nabla v = \overline{\mathbf{B}}^{eT} \mathbf{v}^e$, where $\overline{\mathbf{B}}^e = \left\{ \frac{\partial \overline{\tilde{\phi}_i}^e}{\partial X} \quad \frac{\partial \tilde{\phi}_i^e}{\partial Y} \right\}$.

Substituting these into Eqn. 4.9, we get:

$$\int_{T_i} \left(\overline{\mathbf{B}}^{eT} \mathbf{v}^e \right)^T (Q \, P_0 H^3 \, \nabla P_0 + Q \, P_0 H^3 \, \mathbf{B}^{eT} \partial \mathbf{P}^e + Q \, \boldsymbol{\phi}^{eT} \partial \mathbf{P}^e H^3 \, \nabla P_0) \, \mathrm{d}A$$

$$+ \int_{T_i} \tilde{\boldsymbol{\phi}}^{eT} \mathbf{v}^e \, \boldsymbol{\Lambda} \cdot (H \, \nabla P_0 + P_0 \, \nabla H + H \, \mathbf{B}^{eT} \partial \mathbf{P}^e + \boldsymbol{\phi}^{eT} \partial \mathbf{P}^e \, \nabla H) \, \mathrm{d}A$$

$$- \int_{\partial T_i \backslash \partial \mathcal{S}} \tilde{\boldsymbol{\phi}}^{eT} \mathbf{v}^e (Q \, P_0 H^3 \, \nabla P_0 + Q \, P_0 H^3 \, \mathbf{B}^{eT} \partial \mathbf{P}^e + Q \, \boldsymbol{\phi}^{eT} \partial \mathbf{P}^e H^3 \, \nabla P_0) \cdot \mathbf{n} \, \mathrm{d}S = 0 \tag{4.12}$$

Collecting the terms we can rewrite this as:

$$\mathbf{v}^{eT} \left[\int_{T_i} (QP_0 H^3 \, \overline{\mathbf{B}}^e \mathbf{B}^{eT} + QH^3 \, \overline{\mathbf{B}}^e \nabla P_0 \, \boldsymbol{\phi}^{eT} + H \, \widetilde{\boldsymbol{\phi}}^e \mathbf{\Lambda}^T \mathbf{B}^{eT} + \widetilde{\boldsymbol{\phi}}^e \mathbf{\Lambda}^T \nabla H \, \boldsymbol{\phi}^{eT}) \, \mathrm{dA} \right] \partial \mathbf{P}^e$$

$$+ \mathbf{v}^{eT} \left[\int_{T_i} (QP_0 H^3 \, \overline{\mathbf{B}}^e \nabla P_0 + H \, \widetilde{\boldsymbol{\phi}}^e \mathbf{\Lambda}^T \nabla P_0 + P_0 \widetilde{\boldsymbol{\phi}}^e \mathbf{\Lambda}^T \nabla H) \, \mathrm{dA} \right]$$

$$- \mathbf{v}^{eT} \int_{\partial T_i \backslash \partial \mathcal{S}} \widetilde{\boldsymbol{\phi}}^e \, (Q \, P_0 H^3 \, \nabla P_0 + Q \, P_0 H^3 \, \mathbf{B}^{eT} \partial \mathbf{P}^e + Q \, \boldsymbol{\phi}^{eT} \partial \mathbf{P}^e H^3 \, \nabla P_0) \cdot \mathbf{n} \, \mathrm{dS} = 0 \quad (4.13)$$

Since the test functions are arbitrary, Eqn. 4.12 reduces to:

$$\mathbf{K}^e \, \partial \mathbf{P}^e - \mathbf{R}^e - \int_{\partial T_i \backslash \partial \mathcal{S}} \widetilde{\boldsymbol{\phi}}^e \, (Q \, P_0 H^3 \, \nabla P_0 + Q \, P_0 H^3 \, \mathbf{B}^{eT} \partial \mathbf{P}^e + Q \, \boldsymbol{\phi}^{eT} \partial \mathbf{P}^e H^3 \, \nabla P_0) \cdot \mathbf{n} \, \mathrm{dS} = 0$$

$$(4.14)$$

where $\mathbf{K}^e$ and $\mathbf{R}^e$ are the element stiffness matrix and element flux vector defined as:

$$\mathbf{K}^e = \int_{T_i} (QP_0 H^3 \, \overline{\mathbf{B}}^e \mathbf{B}^{eT} + QH^3 \, \overline{\mathbf{B}}^e \nabla P_0 \, \boldsymbol{\phi}^{eT} + H \, \widetilde{\boldsymbol{\phi}}^e \mathbf{\Lambda}^T \mathbf{B}^{eT} + \widetilde{\boldsymbol{\phi}}^e \mathbf{\Lambda}^T \nabla H \, \boldsymbol{\phi}^{eT}) \, \mathrm{dA} \quad (4.15)$$

$$\mathbf{R}^e = - \int_{T_i} (QP_0 H^3 \, \overline{\mathbf{B}}^e \nabla P_0 + H \, \widetilde{\boldsymbol{\phi}}^e \mathbf{\Lambda}^T \nabla P_0 + P_0 \widetilde{\boldsymbol{\phi}}^e \mathbf{\Lambda}^T \nabla H) \, \mathrm{dA} \quad (4.16)$$

Assembly

In order to obtain the complete pressure profile over the slider $\mathcal{S}$, Eqn. 4.13 needs to be simultaneously solved over all of the elements. Thus the equations are assembled to form the global stiffness matrix and the global flux vector. During assembly, the flux discontinuities between the elements, accounted for by the third term in Eqn. 4.13, are neglected and hence the global system of equations is obtained as:

$$\mathbf{K} \, \partial \mathbf{P} = \mathbf{R} \quad (4.17)$$

where, $\mathbf{K} = \underset{i=1}{\overset{N_e}{\mathbf{A}}} \mathbf{K}^e$, $\partial \mathbf{P} = \underset{i=1}{\overset{N_e}{\mathbf{A}}} \partial \mathbf{P}^e$ and $\mathbf{R} = \underset{i=1}{\overset{N_e}{\mathbf{A}}} \mathbf{R}^e$, and $\mathbf{A}$ is the assembly operator.

Solution

The above method is implemented as an iterative Newton scheme with line search to ensure rapid convergence. The resulting system of equations obtained at each iteration is sparse, but not ordered and has a large bandwidth (see Fig. 4.4). In a previous study Dutto [1993] found that the Reverse Cuthill McKee algorithm gives good orderings which result in rapid convergence using the ILU(0) (Incomplete LU with no fill-in) preconditioned GMRES (Generalized Minimum Residual) methods for solving the compressible Navier Stokes equations. Hence we use the Reverse Cuthill McKee algorithm [see Quarteroni, Sacco, and Saleri, 2000] for bandwidth reduction to obtain a banded matrix (as in Fig. 4.5). The system is then solved using the ILU(0) Preconditioned Restarted GMRES Method [see Saad, 2003] with adaptive subspace selection.

Calculation of forces, moments

Once the pressure field over the slider has been determined, the forces and moments about the pivot location can be determined by simply integrating the pressures over the elements using the following relations:

$$F_z = p_0 L^2 \sum_{j=1}^{N_e} \int_{T_j} P \, dA = \mathbf{C}_{F_z}^T \mathbf{P} \tag{4.18}$$

$$M_{pitch} = p_0 L^3 \sum_{j=1}^{N_e} \int_{T_j} P \cdot X \, dA = \mathbf{C}_{M_{pitch}}^T \mathbf{P} \tag{4.19}$$

$$M_{roll} = p_0 L^3 \sum_{j=1}^{N_e} \int_{T_j} P \cdot Y \, dA = \mathbf{C}_{M_{roll}}^T \mathbf{P} \tag{4.20}$$

where the coefficient vectors $\mathbf{C}_{F_z}$, $\mathbf{C}_{M_{pitch}}$ and $\mathbf{C}_{M_{roll}}$ are defined as:

$$\mathbf{C}_{F_z} = p_0 L^2 \, \mathop{\mathbf{A}}_{i=1}^{N_e} \int_{T_i} \phi^e \, \mathrm{dA} \qquad (4.21)$$

$$\mathbf{C}_{M_{pitch}} = p_0 L^3 \, \mathop{\mathbf{A}}_{i=1}^{N_e} \int_{T_i} X \phi^e \, \mathrm{dA} \qquad (4.22)$$

$$\mathbf{C}_{M_{roll}} = p_0 L^3 \, \mathop{\mathbf{A}}_{i=1}^{N_e} \int_{T_i} Y \phi^e \, \mathrm{dA} \qquad (4.23)$$

4.3.4 Inverse problem

In the inverse problem the net forces and moments on the slider are known and the pressure distribution and the flying attitude of the slider are to be determined. The inverse problem can be solved in two ways, 1) by linearizing Eqn. 4.2 and solving a series of forward problems, or, 2) by linearizing Eqn. 4.8 along with Eqn. 4.2 and solving the resulting system. The first method requires two levels of iterations, the first one to solve the linearized version of Eqn. 4.2 and the second level to solve the forward problem at each of iteration. The second method on the other hand requires just one level of iterations, which however are more difficult to solve and take longer to converge.

4.3.5 Inverse Problem Method 1: Series of Forward Problems

Linearizing the governing equation Eqn. 4.2 about $\mathbf{x}_0$, we get the following equation:

$$\mathbf{K} \, \partial \mathbf{x} - \mathbf{K}_{abs} \, \partial \mathbf{x} + \mathbf{K}_{con} \, \partial \mathbf{x} = \mathbf{F}_{abs}(\mathbf{x}_0) + \mathbf{F}_{con}(\mathbf{x}_0) - \mathbf{K} \, \mathbf{x} - \mathbf{F}_{susp} \qquad (4.24)$$

where $\mathbf{K}_{abs}$ and $\mathbf{K}_{con}$ are the air-bearing and contact stiffnesses of the slider-disk interface.

Equation 4.24 can be formulated in an iterative form and the solution to the resulting problem will involve solving the forward problem (at $\mathbf{x}_0$) along with the air-bearing stiffness matrix. Solution of the forward problem has already been discussed in the previous section.

Air bearing stiffness calculation

The air-bearing stiffness matrix is defined as the change in forces and moments of the air-bearing due to changes in the flying attitude of the slider. To determine these, we need to find the change in pressure P due to changes in the attitude of the slider. The clearance H depends on the attitude of the slider as:

$$H = \frac{1}{h_m} \left(d_{etch} + z_{pivot} + XL \cdot \theta_{pitch} + YL \cdot \theta_{roll} \right) = H(z_{pitch}, \theta_{pitch}, \theta_{roll}) \tag{4.25}$$

where d_{etch} is the etch depth, z_{pivot} is the z-height of the pivot location, θ_{pitch} is the pitch angle and θ_{roll} is the roll angle, and X and Y are the coordinates of the point measured from the pivot location (see Fig. 4.2). We can write the weak form of the time-independent Generalized Reynolds equation, Eqn. 4.5 as:

$$\vartheta(P, H) = 0 \tag{4.26}$$

Differentiating with respect to z_{pivot}, we get:

$$\frac{d\vartheta(P, H)}{dz_{pivot}} = \frac{\partial \vartheta(P, H)}{\partial H} \cdot \frac{\partial H}{\partial z_{pivot}} + \frac{\partial \vartheta(P, H)}{\partial P} \cdot \frac{dP}{dz_{pivot}} = 0 \tag{4.27}$$

$$\Rightarrow \frac{dP}{dz_{pivot}} = \left[\frac{\partial \vartheta(P, H)}{\partial P} \right]^{-1} \left\{ \frac{\partial \vartheta(P, H)}{\partial H} \cdot \frac{\partial H}{\partial z_{pivot}} \right\} \tag{4.28}$$

Similarly differentiating with respect to θ_{pitch} and θ_{roll}, we obtain:

$$\frac{\mathrm{d}P}{\mathrm{d}\theta_{pitch}} = \left[\frac{\partial\vartheta(P,H)}{\partial P}\right]^{-1}\left\{\frac{\partial\vartheta(P,H)}{\partial H}\cdot\frac{\partial H}{\partial\theta_{pitch}}\right\} \tag{4.29}$$

$$\frac{\mathrm{d}P}{\mathrm{d}\theta_{roll}} = \left[\frac{\partial\vartheta(P,H)}{\partial P}\right]^{-1}\left\{\frac{\partial\vartheta(P,H)}{\partial H}\cdot\frac{\partial H}{\partial\theta_{roll}}\right\} \tag{4.30}$$

Substituting the finite element interpolations (Eqns. 4.10, 4.11) and evaluating the expressions above, we get:

$$\left\{\frac{\mathrm{d}\mathbf{P}}{\mathrm{d}z_{pivot}}\right\} = \left[\,\mathbf{K}\,\right]^{-1}\left\{\frac{\partial\mathbf{R}}{\partial z_{pivot}}\right\} \tag{4.31}$$

$$\left\{\frac{\mathrm{d}\mathbf{P}}{\mathrm{d}\theta_{pitch}}\right\} = \left[\,\mathbf{K}\,\right]^{-1}\left\{\frac{\partial\mathbf{R}}{\partial\theta_{pitch}}\right\} \tag{4.32}$$

$$\left\{\frac{\mathrm{d}\mathbf{P}}{\mathrm{d}\theta_{roll}}\right\} = \left[\,\mathbf{K}\,\right]^{-1}\left\{\frac{\partial\mathbf{R}}{\partial\theta_{roll}}\right\} \tag{4.33}$$

where $\mathbf{K}$ is the global stiffness matrix and the vectors $\frac{\partial\mathbf{R}}{\partial z_{pivot}}$, $\frac{\partial\mathbf{R}}{\partial\theta_{pitch}}$ and $\frac{\partial\mathbf{R}}{\partial\theta_{roll}}$ are defined as:

$$\left\{\frac{\partial\mathbf{R}}{\partial z_{pivot}}\right\} = -\frac{1}{h_m}\overset{N_e}{\underset{i=1}{\mathbf{A}}}\int_{T_i}(3\,QP_0H^2\,\overline{\mathbf{B}}^e\nabla P_0 + \widetilde{\phi}^e\mathbf{\Lambda}^T\nabla P_0)\,\mathrm{d}A \tag{4.34}$$

$$\left\{\frac{\partial\mathbf{R}}{\partial\theta_{pitch}}\right\} = -\frac{L}{h_m}\overset{N_e}{\underset{i=1}{\mathbf{A}}}\int_{T_i}(3\,QP_0H^2\,\overline{\mathbf{B}}^e\nabla P_0 + \widetilde{\phi}^e\mathbf{\Lambda}^T\nabla P_0)\cdot X\,\mathrm{d}A \tag{4.35}$$

$$\left\{\frac{\partial\mathbf{R}}{\partial\theta_{roll}}\right\} = -\frac{L}{h_m}\overset{N_e}{\underset{i=1}{\mathbf{A}}}\int_{T_i}(3\,QP_0H^2\,\overline{\mathbf{B}}^e\nabla P_0 + \widetilde{\phi}^e\mathbf{\Lambda}^T\nabla P_0)\cdot Y\,\mathrm{d}A \tag{4.36}$$

The terms of the 3×3 stiffness matrix can then be evaluated as:

$$\mathbf{K}_{abs} = \begin{bmatrix} \mathbf{C}_{F_z}^T\left\{\frac{\partial\mathbf{R}}{\partial z_{pivot}}\right\} & \mathbf{C}_{M_{pitch}}^T\left\{\frac{\partial\mathbf{R}}{\partial z_{pivot}}\right\} & \mathbf{C}_{M_{roll}}^T\left\{\frac{\partial\mathbf{R}}{\partial z_{pivot}}\right\} \\[3mm] \mathbf{C}_{F_z}^T\left\{\frac{\partial\mathbf{R}}{\partial\theta_{pitch}}\right\} & \mathbf{C}_{M_{pitch}}^T\left\{\frac{\partial\mathbf{R}}{\partial\theta_{pitch}}\right\} & \mathbf{C}_{M_{pitch}}^T\left\{\frac{\partial\mathbf{R}}{\partial\theta_{pitch}}\right\} \\[3mm] \mathbf{C}_{F_z}^T\left\{\frac{\partial\mathbf{R}}{\partial\theta_{roll}}\right\} & \mathbf{C}_{M_{pitch}}^T\left\{\frac{\partial\mathbf{R}}{\partial\theta_{roll}}\right\} & \mathbf{C}_{M_{roll}}^T\left\{\frac{\partial\mathbf{R}}{\partial\theta_{roll}}\right\} \end{bmatrix} \tag{4.37}$$

Thus the stiffness is obtained by the solution of three extra linear systems. However this is not computationally very expensive even with iterative methods (where the system matrix **K** has not been factorized) since preconditioners for **K** will already have been evaluated.

Solution

The equation of motion is solved using the Newton Raphson method by repeatedly solving the linearized equation. The forward problem is solved at each iteration. Again we use line search to speed up convergence of the iterations. Stiffness calculations for the air bearing are made after every few iterations (and when line-search fails) with symmetric Broyden updates of rank-2 after each iteration. This avoids wasting too much time on calculating the stiffness matrix at each iteration, while still maintaining a good approximation of the stiffness matrix at each iteration.

4.3.6   Inverse Problem Method 2: Fully linearized system

The second method for solving the inverse problem is to linearize the governing equation Eqn. 4.2 and the Reynolds equation together, and then solve only one level of iterations.

Linearization

Consider again the elemental weak form of the Generalized Reynolds equation, Eqn. 4.8. Linearizing with respect to $\{P, H\}$ about $\{P_0, H_0\}$, we get:

$$\int_{T_i} \nabla v \cdot (Q\, P_0 H_0^3\, \nabla P_0 + Q\, P_0 H_0^3\, \nabla \partial P + Q\, \partial P H_0^3\, \nabla P_0 + 3Q\, P_0 H_0^2\, \partial H\, \nabla P_0)\, \mathrm{d}A$$

$$+ \int_{T_i} v\, \mathbf{\Lambda} \cdot (H_0\, \nabla P_0 + P_0\, \nabla H_0 + H_0\, \nabla \partial P + \partial P\, \nabla H_0 + \partial H\, \nabla P_0 + P_0\, \nabla \partial H)\, \mathrm{d}A$$

$$-\!\!\!\int_{\partial T_i \backslash \partial S} v\, (Q\, P_0 H_0^3\, \nabla P_0 + Q\, P_0 H_0^3\, \nabla \partial P + Q\, \partial P H_0^3\, \nabla P_0 + 3Q\, P_0 H_0^2\, \partial H\, \nabla P_0) \cdot \mathbf{n}\, \mathrm{d}S = 0$$

$$(4.38)$$

The above equation is referred to as the 'fully linearized' form of the elemental weak form.

Finite element discretization

Substituting the discrete forms of the pressure (Eqn. 4.10) and the test functions (Eqn. 4.11), we obtain:

$$\int_{T_i} \left(\overline{\mathbf{B}}^{eT} \mathbf{v}^e\right)^T (Q\, P_0 H_0^3\, \nabla P_0 + Q\, P_0 H_0^3\, \mathbf{B}^{eT} \partial \mathbf{P}^e + Q\, \boldsymbol{\phi}^{eT} \partial \mathbf{P}^e H_0^3\, \nabla P_0 + 3Q\, P_0 H_0^2\, \partial H\, \nabla P_0)\, \mathrm{d}A$$

$$+ \int_{T_i} \widetilde{\boldsymbol{\phi}}^{eT} \mathbf{v}^e\, \mathbf{\Lambda} \cdot (H_0\, \nabla P_0 + P_0\, \nabla H_0 + H_0\, \mathbf{B}^{eT} \partial \mathbf{P}^e + \boldsymbol{\phi}^{eT} \partial \mathbf{P}^e\, \nabla H + \partial H\, \nabla P_0 + P_0\, \nabla \partial H_0)\, \mathrm{d}A$$

$$-\!\!\!\int_{\partial T_i \backslash \partial S} \widetilde{\boldsymbol{\phi}}^{eT} \mathbf{v}^e\, (Q\, P_0 H_0^3\, \nabla P_0 + Q\, P_0 H_0^3\, \mathbf{B}^{eT} \partial \mathbf{P}^e + Q\, \boldsymbol{\phi}^{eT} \partial \mathbf{P}^e H_0^3\, \nabla P_0 + 3Q\, P_0 H_0^2\, \partial H_0\, \nabla P_0) \cdot \mathbf{n}\, \mathrm{d}S = 0$$

$$(4.39)$$

Now using the definition of H in Eqn. 4.25, we can rewrite ∂H and $\nabla \partial H$ as:

$$\partial H = \frac{1}{h_m}\left(\partial z_{pivot} + XL \cdot \partial \theta_{pitch} + YL \cdot \partial \theta_{roll}\right) = \mathbf{C}_H^T \partial \mathbf{x} \qquad (4.40)$$

$$\nabla \partial H = \frac{L}{h_m}\left(\partial \theta_{pitch}\, \mathbf{E}_X + \partial \theta_{roll}\, \mathbf{E}_Y\right) = \mathbf{C}_{\nabla H}^T \partial \mathbf{x} \qquad (4.41)$$

where $\mathbf{C}_H = \frac{1}{h_m} \left\{ \begin{array}{ccc} 1 & XL & YL \end{array} \right\}^T$ and $\mathbf{x} = \left\{ \begin{array}{ccc} z_{pivot} & \theta_{pitch} & \theta_{roll} \end{array} \right\}^T$ and $\mathbf{C}_{\nabla H} = \frac{L}{h_m} \left[\begin{array}{ccc} 0 & 1 & 0 \\ 0 & 0 & 1 \end{array} \right]^T$

Thus we get:

$$\mathbf{K}^e_{full} \left\{ \begin{array}{cc} \partial \mathbf{P}^{eT} & \partial \mathbf{x}^T \end{array} \right\}^T - \mathbf{R}^e_{full} -$$

$$\int_{\partial T_i \backslash \partial S} \widetilde{\phi}^e \; (Q \; P_0 H_0^3 \; \nabla P_0 + Q \; P_0 H_0^3 \; \mathbf{B}^{eT} \partial \mathbf{P}^e + Q \; \phi^{eT} \partial \mathbf{P}^e H_0^3 \; \nabla P_0 + 3Q \; P_0 H_0^2 \; \partial H \; \nabla P_0) \cdot \mathbf{n} \; \mathrm{dS} = 0$$

$$(4.42)$$

where $\mathbf{K}^e_{full}$ and $\mathbf{R}^e_{full}$ are the element stiffness matrix and element flux vector defined as:

$$\mathbf{K}^e_{full} = \left[\begin{array}{cc} \mathbf{K}^e_P & \mathbf{K}^e_H \end{array} \right] \tag{4.43}$$

$$\mathbf{K}^e_P = \int_{T_i} (Q \; P_0 H_0^3 \; \overline{\mathbf{B}}^e \mathbf{B}^{eT} + Q \; H_0^3 \; \overline{\mathbf{B}}^e \nabla P_0 \; \phi^{eT} + H_0 \; \widetilde{\phi}^e \mathbf{\Lambda}^T \mathbf{B}^{eT} + \widetilde{\phi}^e \mathbf{\Lambda}^T \nabla H \; \phi^{eT}) \; \mathrm{dA}$$

$$\mathbf{K}^e_H = \int_{T_i} (3Q \; P_0 H_0^2 \; \overline{\mathbf{B}}^e \mathbf{B}^{eT} \nabla P_0 \; \mathbf{C}_H + \widetilde{\phi}^e \mathbf{\Lambda}^T \nabla P_0 \; \mathbf{C}_H + P_0 \; \widetilde{\phi}^e \mathbf{\Lambda}^T \; \mathbf{C}_{\nabla H}) \; \mathrm{dA}$$

$$\mathbf{R}^e_{full} = - \int_{T_i} (QP_0 H_0^3 \; \overline{\mathbf{B}}^e \nabla P_0 + H_0 \; \widetilde{\phi}^e \mathbf{\Lambda}^T \nabla P_0 + P_0 \widetilde{\phi}^e \mathbf{\Lambda}^T \nabla H_0) \; \mathrm{dA} \tag{4.44}$$

Assembly

After assembling the set of equations in Eqn. 4.42, we get a system of N_n equations in $N_n + 3$ unknowns. We need 3 more equations to determine the solution of the system. These three equations are the linearized governing equations (Eqn. 4.24) in the z_{pivot}, θ_{pitch} and θ_{roll} directions. Thus we end up with the following system of equations:

$$\left[\begin{array}{cc} \mathbf{K}_P & \mathbf{K}_H \\ \mathbf{C}_{abs} & \mathbf{K}_{eff} \end{array} \right] \left\{ \begin{array}{c} \partial \mathbf{P} \\ \partial \mathbf{x} \end{array} \right\} = \left\{ \begin{array}{c} \mathbf{R}_{full} \\ \mathbf{R}_H \end{array} \right\} \tag{4.45}$$

where:

$$\mathbf{K}_P = \mathop{\mathbf{A}}_{i=1}^{N_e} \mathbf{K}_P^e \qquad (4.46)$$

$$\mathbf{K}_H = \mathop{\mathbf{A}}_{i=1}^{N_e} \mathbf{K}_H^e \qquad (4.47)$$

$$\partial \mathbf{P} = \mathop{\mathbf{A}}_{i=1}^{N_e} \partial \mathbf{P}^e \qquad (4.48)$$

$$\mathbf{R}_{full} = \mathop{\mathbf{A}}_{i=1}^{N_e} \mathbf{R}_{full}^e \qquad (4.49)$$

$$\mathbf{C}_{abs} = \left\{ \begin{array}{ccc} \mathbf{C}_{F_z} & \mathbf{C}_{M_{pitch}} & \mathbf{C}_{M_{roll}} \end{array} \right\} \qquad (4.50)$$

$$\mathbf{K}_{eff} = \mathbf{K}_{susp} + \mathbf{K}_{abs} + \mathbf{K}_{con} \qquad (4.51)$$

$$\mathbf{R}_H = \mathbf{C}_{abs}^T \mathbf{P}_0 + \mathbf{F}_{con}(\mathbf{x}_0) - \mathbf{K}_{susp}\mathbf{x}_0 - \mathbf{F}_{susp} \qquad (4.52)$$

Solution

The formulation described above is again implemented using a Newton Raphson iterative scheme. The resulting system of equations at each iteration is also sparse, however the equations corresponding to the balance of forces/moments are full and the coefficients corresponding to the slider attitude are nonzero for all of the equations. Hence after the application of the Reverse Cuthill McKee algorithm, we end up with an 'arrowhead' matrix (see Fig. 4.6). For simplicity, the arrowhead matrix is also solved using the preconditioned GMRES algorithm used for the forward problem and the inverse problem solved using method 1. The method is found to be slower than method 1 using this algorithm, however it is expected to be faster using either a direct method or a preconditioner which exploits the structure of the matrix.

4.3.7   Streamline Upwind/Petrov-Galerkin formulation

The finite element method makes use of a weighted residual formulation to arrive at a system of equations. The most common type of weighting used is the Galerkin method, wherein the basis functions for the pressure P and the test functions v are the same. However, it is observed that when the Galerkin method is applied to convection dominated convective-diffusive equations, the system matrices are non-symmetric due to the convection terms. This leads to a loss of the 'best approximation' property and the resulting solutions are often corrupted by spurious oscillations (see Fig. 4.7(a)). One way to eliminate these oscillations is by refining the mesh such that convection no longer dominates on the element level. But this refinement is needlessly expensive. It was discovered that the spurious oscillations can be removed by using 'upwind' methods which amounts to adding artificial diffusion in the streamline direction. Probably the most successful of the upwinding schemes is the Streamline upwind/Petrov-Galerkin (SUPG) formulation presented by Brooks and Hughes [1982]. They present a formulation in which the standard Galerkin functions are modified by adding a streamline upwind perturbation. We shall use the SUPG formulation for our finite element method. With the basis $\tilde{\phi}$ for our test function v is modified to:

$$\tilde{\phi} = \phi + \tilde{k}\,\frac{\mathbf{u}\cdot\nabla\phi}{||\mathbf{u}||^2} \tag{4.53}$$

and where for our problem $\mathbf{u}$ and $\tilde{k}$ are defined as,

$$\mathbf{u} = H\boldsymbol{\Lambda} - 2Q\ H^3\ \nabla P - 3Q\ PH^2\ \nabla H \tag{4.54}$$

$$\tilde{k} = \frac{||\mathbf{u}||\ h}{2}\ \tilde{\xi} \tag{4.55}$$

$$\tilde{\xi} = \coth(\alpha) - \frac{1}{\alpha} \tag{4.56}$$

$$\alpha = \frac{||\mathbf{u}||\ h}{2Q\ H^3 P} \tag{4.57}$$

where α is the Peclet number and h is the characteristic element length as shown in Fig. 4.8.

4.3.8 The force norm

For the analysis to be carried out in the subsequent sections of this chapter, we define a new norm: the *force* norm. The force norm is a weighted norm that can be directly used to correlate fields (pressures etc.) to their integrals over the domain $\mathcal{S}$ (forces etc.). The force norm is defined as:

$$|\zeta|_F = p_0 L^2 \sum_{j=1}^{N_e} \int_{T_j} |\zeta_i|\ \phi_i^e\ \mathrm{d}A \approx \mathbf{C}_{F_z}^T\ |\zeta| \tag{4.58}$$

4.4 Numerical Simulations

4.4.1 Meshing

Mesh Generation

Meshes for the sliders are generated using *TRIANGLE*, a robust 2D triangular mesh generator developed by Shewchuk [1996]. In order to ensure that errors in the solutions are small, the mesh is automatically refined based on certain criteria have been proposed and tested. Refinement and regeneration of the meshes is also done using *TRIANGLE*. The coarsest level meshes are generated such that the element edges conform to the slider's rail and wall geometries. Coarse level meshes for two slider designs are shown in Fig. 4.9. Various criteria for refinement are investigated in section 4.4.4.

Solution prolongation

Our formulation is based on the Newton Raphson scheme, which requires an initial guess of the pressure for the solution. For the coarsest mesh, we use ambient pressure as the initial guess. However when meshes are refined, using ambient pressure as the initial guess is wasteful because we already have the converged solution on the coarse mesh. Hence the solution is projected from the coarse mesh onto the fine mesh. Since the solution is only an initial guess for the finer mesh, we use simple linear interpolation over the coarse elements to determine nodal pressure values over the fine mesh. The problem then reduces to determining which element of the coarse mesh contains any given node of the new mesh. This is accomplished by a greedy algorithm which moves across adjoining elements using the distance between the fine mesh node and the coarse element centroid as a cost function. This algorithm is simple and fast and a solution is guaranteed since in the worst case all of the triangles on the coarse mesh will be traversed.

4.4.2 Implementation

The formulations described above were implemented in C++ and studies were carried out to investigate grid convergence. The code is available for CML members to download from the Computer Mechanics Lab website (*http://cml.me.berkeley.edu*).

4.4.3 Forward problem

Three slider designs are simulated for increasing grid sizes. These are the flat slider, the 2-rail slider and Slider 1 shown in Figs. 4.10, 4.11 and 4.12. The operating attitudes of the three designs are listed in Table 4.2. The grid used for each of the meshes is a uniform triangular grid.

Figure 4.13 shows plots of the resultant pressure force and pitch-moment for the flat slider. In Fig. 4.14, we plot the error at each level. The error is computed by calculating the pressure profile at a very fine grid and treating it as the exact solution. Data points for fine meshes are neglected since they appear to be too close to the mesh used for calculating the error. We see that the order of accuracy is close to 1.1 for both the force and the pitch-moment. The pressure profile for the flat slider can be seen in Fig. 4.15, where we observe a very large pressure gradient near the trailing edge of the slider which increases the interpolation error in the solution, causing the solution to have a low order of convergence. Figure 4.16 shows plots of the resultant pressure force and pitch-moment for the 2-rail slider. In Fig. 4.17, we plot the error at each level. The pressure profile for the 2-rail slider can be seen in Fig. 4.18. Again we see a large pressure gradient near the trailing edge of the slider however this is restricted to one corner of the slider. The rest of the pressure profile appears smooth. Hence an order of convergence of 1.2 is observed numerically. Figure 4.19 shows plots of the resultant pressure force and pitch-moment for slider 1. In Fig. 4.20, we plot the error at each level. The pressure profile for slider 1 can be seen in Fig. 4.21. No large pressure gradients are observed and hence an order of convergence of 1.5 is observed numerically.

Table 4.2: Slider attitudes for grid convergence studies

	Flat	2-Rail	Slider 1
Nominal FH (nm)	15	15	20
Pitch (μrad)	150	150	200
Roll (μrad)	0	0	1
Radius (mm)	23	23	32
Skew ($^\circ$)	-9.1	-9.1	0.0
RPM	10K	10K	10K

Thus we see that the order of convergence of the solution for our problem is strongly dependent on the slider geometry and the operating parameters of the slider. However the grid convergence discussed in this section is for a uniform mesh. We can artificially enhance grid convergence by using selective mesh refinement to refine only areas of the slider where the error is high. This will be discussed in the next sub-section.

4.4.4 Refinement strategies

In this section we will study various refinement strategies for the mesh so as to obtain the most accurate solution with the fewest number of nodes. We will use the slider design shown in Fig. 4.22 to show how the error changes with various meshing strategies. The error is calculated by assuming that the solution calculated on a very fine mesh is the exact solution. The operating parameters for the slider are tabulated in Table 4.3 and the pressure profile under the slider is plotted in Fig. 4.23. We use the same coarse level mesh for all the refinement cases. The coarse level mesh is a uniform mesh with a maximum element area of 5×10^{-4} (nondimensionalized with respect to L^2). The error in the coarsest mesh and the associated mesh are shown in Figs. 4.24 and 4.25 respectively. In the following sections we will present briefly a study of various refinement strategies. Although there are many parameters that can be adjusted to control refinement, such as refinement thresholds and refinement factors, we will only present representative cases for each refinement strategy.

Table 4.3: Slider parameters for mesh refinement study

	Refinement slider
Nominal FH (nm)	6
Pitch (μrad)	150
Roll (μrad)	0
Radius (mm)	23
Skew ($°$)	-9.1
RPM	10K

Pressure based refinement

The value of the steady state pressure can be used as a refinement criterion to refine the coarse level mesh. This leads to the mesh being refined in areas with higher pressures. The rationale behind using this refinement scheme is that areas with higher pressures contribute more to the forces and moments and hence need to be resolved more accurately than regions with lower pressures.

$$\theta_{pres} = \int_{T_i} |P| \, \mathrm{d}A \qquad (4.59)$$

The errors obtained by using this refinement technique are plotted in Fig. 4.26 for two levels of refinement. The associated meshes are plotted in Fig. 4.27.

Clearance based refinement

The clearance under the slider at the steady state can be used as a refinement criterion to refine the coarse level mesh. Thus areas which are closer to the disk (rails and areas closer to the trailing edge) will have a finer mesh than other areas. The rationale here is that areas closer to the disk typically have high pressure and may have higher errors.

$$\theta_{spacing} = \frac{1}{\int_{T_i} H \, \mathrm{dA}} \tag{4.60}$$

The estimated errors when using this technique for refinement are plotted in Fig. 4.28 for two levels of refinement. The associated meshes are plotted in Fig. 4.29.

Pressure gradient based refinement

The pressure gradient under the slider at the steady state can be used as a refinement criterion to refine the coarse level mesh. Thus areas which have large pressure gradients (typically near the trailing edge pads on the slider) will have a finer mesh than other areas.

$$\theta_{\nabla P} = \int_{T_i} |\nabla P| \, \mathrm{dA} \tag{4.61}$$

The estimated errors when using this technique for mesh refinement are plotted in Fig. 4.30 for two levels of refinement. The associated meshes are plotted in Fig. 4.31.

Flux jump based refinement

The pressure flux jumps under the slider at the steady state can also be used as a refinement criterion to refine the coarse level mesh. In some sense this measure is representative of the interpolation error of the solution. Thus areas which have large pressure flux discontinuities will have a finer mesh than other areas. We observe that the flux jump terms across the elements were neglected during the assembly of the system equations and hence a larger flux jump term may be an indicator of larger error.

$$\theta_{\|\nabla P\|} = \left| h_i \left(\nabla P^+ - \nabla P^- \right) \cdot \mathbf{n} \right| \tag{4.62}$$

The estimated errors using this technique are plotted in Fig. 4.32 for two levels of refinement. The associated meshes are plotted in Fig. 4.33.

Streamwise clearance gradient based refinement

Next we can also use the streamwise clearance gradient of the slider at the steady state as a refinement criterion to refine the coarse level mesh. Thus areas which have steep clearance gradients (such as on the walls) will have a finer mesh than other areas.

$$\theta_{\nabla H \cdot \lambda} = \int_{T_i} \frac{|\mathbf{\Lambda} \cdot \nabla H|}{|\mathbf{\Lambda}|} \, \mathrm{d}A \tag{4.63}$$

The estimated errors using this technique are plotted in Fig. 4.34 for two levels of refinement. The associated meshes are plotted in Fig. 4.35.

Residual

Finally we explore the strategy wherein the residual of the nonlinear Reynolds equation is used as a refinement criterion to refine the coarse level mesh. Thus areas with higher residuals are used as an indication of higher error and to generate refined meshes.

$$\theta_{R_e} = \int_{T_i} |R_{n+1}^e| \, \mathrm{d}A \tag{4.64}$$

The estimated errors using this technique are plotted in Fig. 4.36 for two levels of refinement. The associated meshes are plotted in Fig. 4.37.

The grid convergence for the refinement slider (shown in Fig. 4.22 for each of the refinement strategies discussed above is plotted in Fig. 4.38. We see that convergence is most rapid when the pressure gradient and the flux jumps are used as refinement criteria. Convergence is slower when the residual and streamwise clearance gradient are used as criteria. Convergence is slowest when pressure is used as a refinement criterion. However when the clearance is used as a refinement criterion, the solution does not converge to the correct value for the grid sizes simulated. This is because errors towards the leading edge of the slider do not get reduced quickly enough.

In Figs. 4.39 and 4.40, we plot the grid convergence for slider 1 (shown in Fig. 4.12) and slider 2 (shown in Fig. 4.41), respectively. A similar convergence behavior is observed as for the refinement slider, with convergence being fastest when pressure gradients and flux jumps are used as the refinement criteria. Looking at the error plots in Figs. 4.26-4.36, we observe that most of the error is concentrated in regions of high pressure gradients, which is often observed in areas with steep clearance gradients, such as along walls of rails, and more so towards the trailing edge. With further investigation we find that convergence can be enhanced by generating an initial mesh which has smaller elements near the walls as compared to on top of the rails. This concept will be further investigated in the next section.

4.4.5 Inverse problem

Three slider designs are simulated for the inverse problem. These are Slider 1, shown in Fig. 4.12, Slider 2 is shown in Fig. 4.41 and Slider 3 which is shown in Fig. 4.42. The operating parameters of the three designs are listed in Table 4.4. The initial grid used for each of the meshes is obtained by the dense-edge strategy described in the subsection on refinement to ensure rapid convergence, wherein the refinement after each level is uniform. The flying attitude, i.e. the nominal fly-height, the pitch and the roll are plotted for each case.

Figure. 4.43 shows plots of the convergence of the slider fly height, pitch and the roll with reducing grid sizes. We see that the result obtained from the coarsest mesh is within half a nanometer of the final converged value. Similarly the pitch is found to be within 2 μrad and roll within 0.2 μrad. In Fig. 4.44 we show plots of the three mesh levels used for the convergence study. Similarly attitude convergence plots for the inverse problem are obtained for sliders 2 and 3 in Figs. 4.45 and 4.46, respectively. For slider 2, the nominal FH is within 0.2 nm and the pitch and roll are within 2 μrad and 0.2 μrad respectively. For slider 3, the nominal FH is within 0.2 nm and the pitch and roll are within 3 μrad and 0.3 μrad, respectively.

4.5 Discussion

A novel method of solving the inverse problem for air-bearing sliders in hard disk drives is presented. The formulation is implemented and convergence studies are carried out for the method. The order of convergence is found to depend on the geometry of the slider and the resulting pressure profile that develops under it. Various strategies for refinement of the mesh are discussed. Refinements based on flux jumps and pressure gradients are found to work well. It is also found that the number of levels required for convergence can be reduced if the initial mesh is generated so as to have a fine mesh along rail walls.

Table 4.4: Slider operation parameters for grid convergence studies

	Slider 1	Slider 2	Slider 3
Gram load (g)	1.5	2.0	3.0
PSA (rad)	0.1	0.0	0.0
RSA (rad)	0.0	0.0	0.0
Radius (mm)	32	32	27
Skew (°)	0.0	3.2	-17.5
RPM	10K	7.2K	7.2K

4.6   Figures

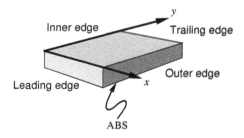

Figure 4.1: Mesh Coordinate System

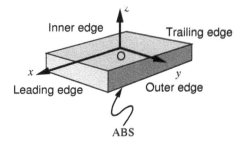

Figure 4.2: System Coordinate System

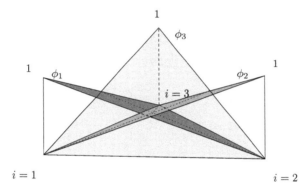

Figure 4.3: Basis Functions

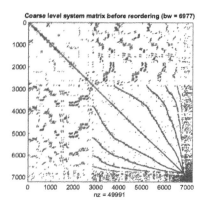

Figure 4.4: Coarse level system matrix before node reordering for the forward problem

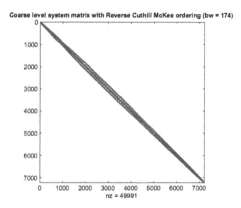

Figure 4.5: Coarse level system matrix after RCM ordering for Method 1

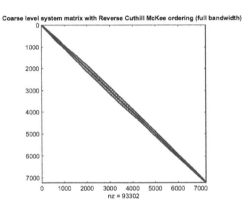

Figure 4.6: Coarse level system matrix after RCM ordering for Method 2 (Note: the last 3 rows and columns are full)

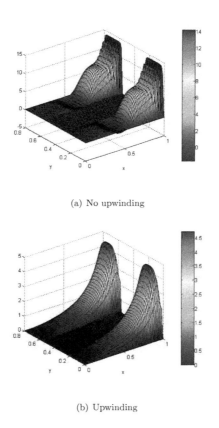

(a) No upwinding

(b) Upwinding

Figure 4.7: Pressure solution for a 2-rail taper slider with and without upwinding

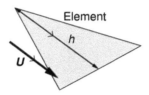

Figure 4.8: Element length scale h

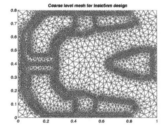 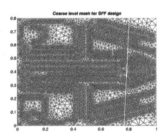

Figure 4.9: Coarse level meshes for two slider designs

Figure 4.10: Flat slider

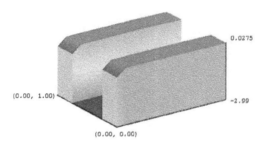

Figure 4.11: 2-rail slider profile

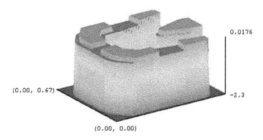

Figure 4.12: Slider 1

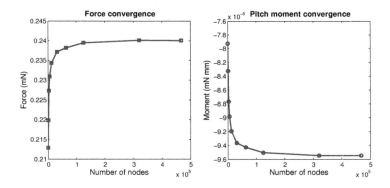

Figure 4.13: Grid convergence for flat slider

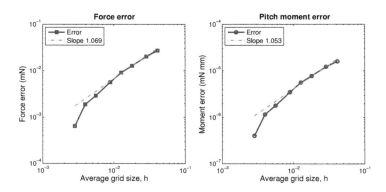

Figure 4.14: Force error for flat slider

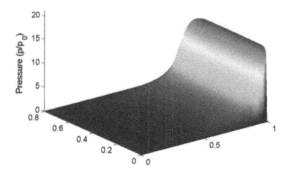

Figure 4.15: Converged pressure profile for flat slider

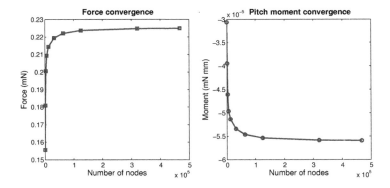

Figure 4.16: Grid convergence for 2-rail slider

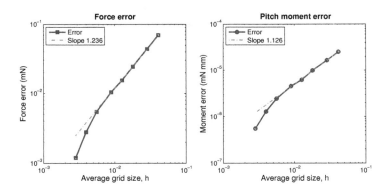

Figure 4.17: Force error for 2-rail slider

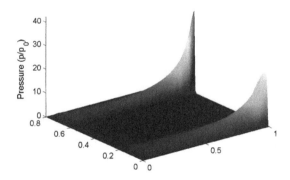

Figure 4.18: Converged pressure profile for 2-rail slider

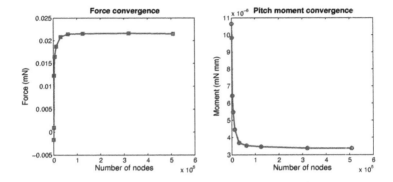

Figure 4.19: Grid convergence for slider 1

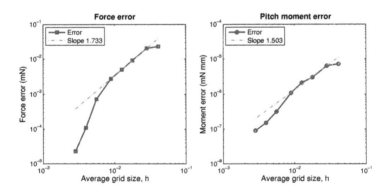

Figure 4.20: Force error for slider 1

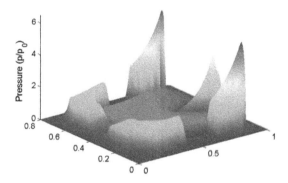

Figure 4.21: Converged pressure profile for slider 1

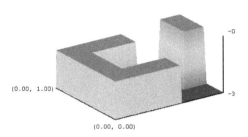

Figure 4.22: Slider design used for refinement studies

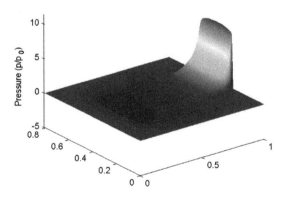

Figure 4.23: Refinement slider pressure profile

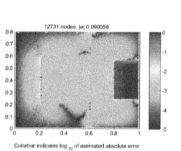

Figure 4.24: Coarse mesh error for refinement slider

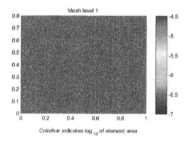

Figure 4.25: Coarse mesh for refinement slider

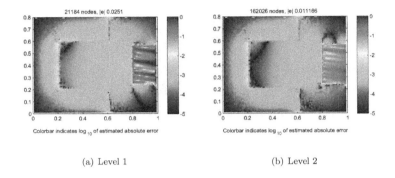

(a) Level 1   (b) Level 2

Figure 4.26: Estimated solution error using pressure based mesh refinement

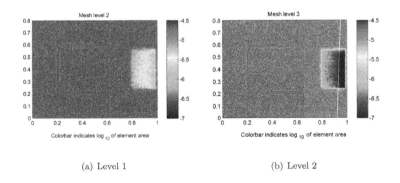

(a) Level 1   (b) Level 2

Figure 4.27: Refined mesh generated using pressure based mesh refinement

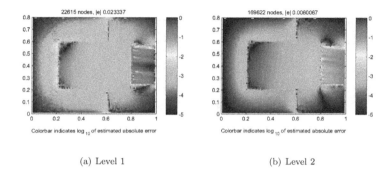

(a) Level 1   (b) Level 2

Figure 4.28: Estimated solution error using clearance based mesh refinement

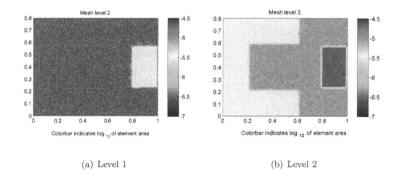

(a) Level 1   (b) Level 2

Figure 4.29: Refined mesh generated using clearance based mesh refinement

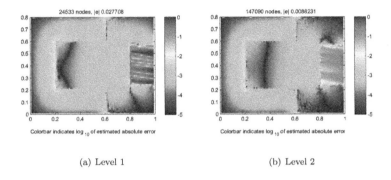

(a) Level 1   (b) Level 2

Figure 4.30: Estimated solution error using pressure gradient based mesh refinement

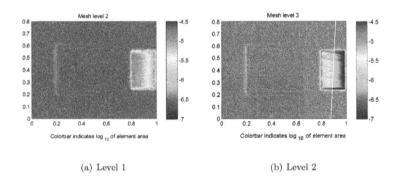

(a) Level 1   (b) Level 2

Figure 4.31: Refined mesh generated using pressure gradient based mesh refinement

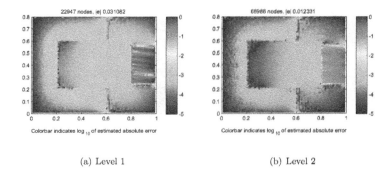

(a) Level 1   (b) Level 2

Figure 4.32: Estimated solution error using flux jump based mesh refinement

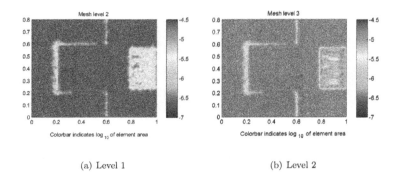

(a) Level 1   (b) Level 2

Figure 4.33: Refined mesh generated using flux jump based mesh refinement

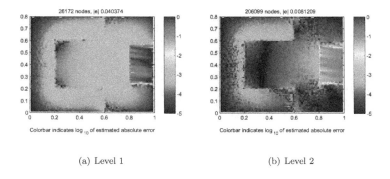

(a) Level 1   (b) Level 2

Figure 4.34: Estimated solution error using clearance gradient based mesh refinement

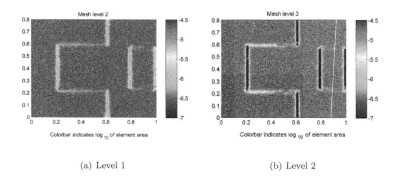

(a) Level 1   (b) Level 2

Figure 4.35: Refined mesh generated using clearance gradient based mesh refinement

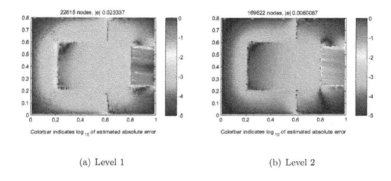

(a) Level 1 (b) Level 2

Figure 4.36: Estimated solution error using clearance gradient based mesh refinement

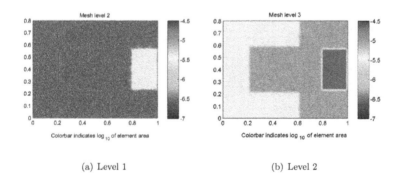

(a) Level 1 (b) Level 2

Figure 4.37: Refined mesh generated using RE residual based mesh refinement

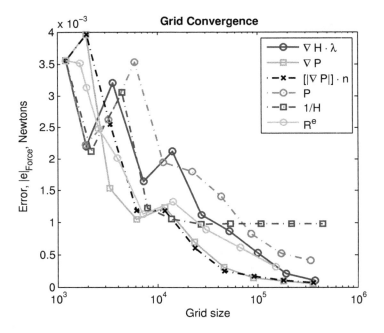

Figure 4.38: Grid convergence for refinement slider using various refinement strategies

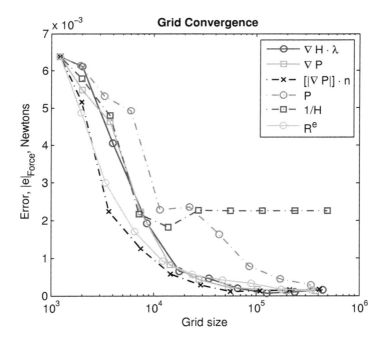

Figure 4.39: Grid convergence for slider 1 using various refinement strategies

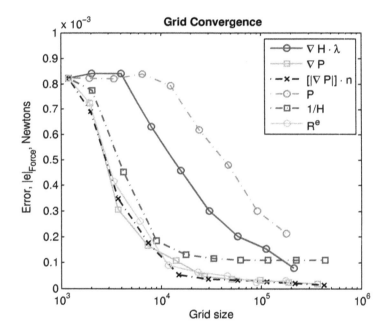

Figure 4.40: Grid convergence for slider 2 using various refinement strategies

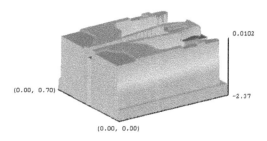

Figure 4.41: Slider 2 profile

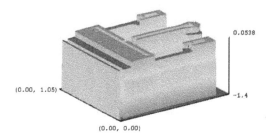

Figure 4.42: Slider 3 profile

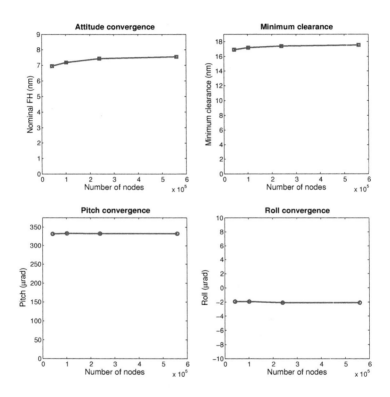

Figure 4.43: Grid convergence for slider 1

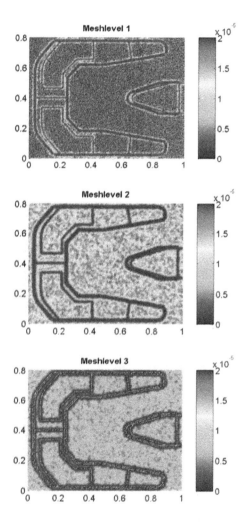

Colorbar indicates normalized area of elements

Figure 4.44: Mesh levels for slider 1

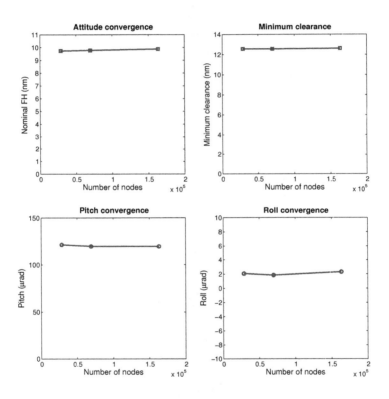

Figure 4.45: Grid convergence for slider 2

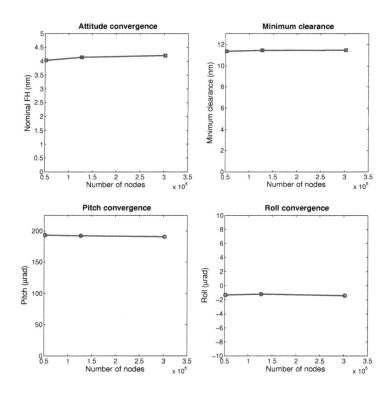

Figure 4.46: Grid convergence for slider 3

Chapter 5

Fluid dynamic modeling

5.1 Introduction

Air bearing designs today are becoming increasingly complex due to requirements of extremely low fly heights and increased reliability in hostile environments. In order to evaluate such designs, there is need for fast, reliable simulations of their dynamic flying characteristics during events such as shock, vibration and load/unload.

In this chapter, we propose and implement a new unstructured finite element based scheme to solve the time-dependent Reynolds equation.

5.2 Previous Work

A good summary of previous work has been presented by Gupta [2007]. Many researchers in the past have made an attempt to predict the system dynamics using a simple head-disk interface model [Ponnaganti, 1986; Lu, 1997] which includes a simplified slider model and an air-bearing model which solves the generalized Reynolds equation. These models ignore the dynamic effects of the suspension and the disk. According to Gupta [2007], it has been shown that the system dynamics predicted by these models is significantly different from the actual system response (measured experimentally) during slider-disk contact/impact, aerodynamic forcing on the HSA due to disk rotation, shocks, track-seek and load-unload. Bhargava and Bogy [2007b] used a method whereby structural simulations carried out in ANSYS, which is a commercial FE Analysis software, were coupled with air bearing simulations obtained by using the finite volume method developed by Hu and Bogy [1995] to simulate the shock response of the suspension/air-bearing/disk system. However this method suffered from overheads relating to data exchange between ANSYS and air bearing calculations. This method was subsequently improved to include suspension and disk models in the form of reduced mass/stiffness matrices which were used to carry out studies on load/unload [Bhargava and Bogy, 2005], shock [Bhargava and Bogy, 2007c] and aerodynamic forcing Gupta [2007]. In this section, we improve upon this method by incorporating the FE scheme developed in Chapter 4 as well as a variable time-stepping scheme based on apparent frequencies.

5.3 Methodology

5.3.1 The generalized Reynolds Equation

The air-bearing pressure used to determine the air-bearing forces and moments is calculated by solving the generalized Reynolds Equation. In this section we present the SUPG (Streamline Upwind/Petrov Galerkin) formulation for the generalized Reynolds

Equation. As discussed in Chapter 4, the spacing between the slider and the disk is ex-

tremely small (much less than the mean free path of air). Under these conditions, the con-

tinuum assumption for the air and the no-slip boundary conditions are no longer valid and

the Reynolds equation is generalized to include rarefaction and slip effects. The generalized

Reynolds Equation can be written in the following form in terms of dimensionless vari-

ables, $P = p/p_0$ (pressure normalized with respect to the ambient pressure p_0), $H = h/h_m$

(slider-disk clearance normalized with respect to a nominal spacing h_m) and $T = \omega \cdot t$ (time

non-dimensionalized with respect to the angular velocity of the disk, ω):

$$\nabla \cdot (Q \, PH^3 \, \nabla P) = \Lambda \cdot \nabla(PH) + \tau \frac{\partial}{\partial T}(PH) \tag{5.1}$$

over the domain of the slider ($\mathcal{S}$) along with the boundary condition of ambient pressure

($P = 1$) at the boundary of the slider ($\partial \mathcal{S}$). In the above equation, $\nabla = \frac{\partial}{\partial X}\mathbf{E}_X + \frac{\partial}{\partial Y}\mathbf{E}_Y$

is the gradient operator with respect to normalized coordinates $X = x/L$ and $Y = y/L$

where L is the characteristic length scale for the slider. The non-dimensional vector Λ is

the bearing number defined as $\Lambda = \frac{6\mu \mathbf{U}L}{p_0 h_m^2}$, where μ is the dynamic viscosity of air and $\mathbf{U}$ is

the local velocity vector of the disk. The squeeze number τ is defined as $\tau = \frac{12\mu\omega L^2}{p_0 h_m^2}$ and

is the ratio of transient effects to the diffusion effects in the problem. The flow factor, Q,

is the modification to the continuum Generalized Reynolds Equation for incorporating slip

and rarefaction effects.

5.3.2 The Weak Form

To derive the weak form of the dynamic generalized Reynolds Equation, Eqn. 5.1, we multiply the equation by a test function v, integrate over the domain $\mathcal{S}$ and use the divergence theorem. Thus we get:

$$\int_{\mathcal{S}} \nabla v \cdot (Q\, PH^3\, \nabla P)\, \mathrm{dA} + \int_{\mathcal{S}} v\, \mathbf{\Lambda} \cdot \nabla(PH)\, \mathrm{dA} + \int_{\mathcal{S}} v\, \tau \frac{\partial}{\partial T}(PH)\, \mathrm{dA}$$
$$- \int_{\partial \mathcal{S}} v\, (Q\, PH^3\, \nabla P) \cdot \mathbf{n}\, \mathrm{dS} = 0 \quad (5.2)$$

The above equation along with the boundary condition $P = 1$ over $\partial \mathcal{S}$ is the weak form of Eqn. 5.1, and it can be solved to give the pressure field over the slider $\mathcal{S}$. We decompose the domain of the slider $\mathcal{S}$ into a finite number of triangular domains, $\mathcal{T}_i$, the *finite elements*, such that:

$$\mathcal{S} = \bigcup_{i=1}^{N_e} \mathcal{T}_i \quad || \quad \mathcal{T}_i \cap \mathcal{T}_j = \emptyset \text{ for } i \neq j \quad (5.3)$$

Writing the weak form Eqn. 5.2 over each element $\mathcal{T}_i$, we have:

$$\int_{\mathcal{T}_i} \nabla v \cdot (Q\, PH^3\, \nabla P)\, \mathrm{dA} + \int_{\mathcal{T}_i} v\, \mathbf{\Lambda} \cdot \nabla(PH)\, \mathrm{dA} + \int_{\mathcal{T}_i} v\, \tau \frac{\partial}{\partial T}(PH)\, \mathrm{dA}$$
$$- \int_{\partial \mathcal{T}_i} v\, (Q\, PH^3\, \nabla P) \cdot \mathbf{n}\, \mathrm{dS} = 0 \quad (5.4)$$

Now since pressure boundary conditions are applied to the boundary of the slider (i.e. pressure is known on $\partial \mathcal{S}$), we set $v = 0$ over $\partial \mathcal{S}$. Thus we have:

$$\int_{\mathcal{T}_i} \nabla v \cdot (Q\, PH^3\, \nabla P)\, \mathrm{dA} + \int_{\mathcal{T}_i} v\, \mathbf{\Lambda} \cdot \nabla(PH)\, \mathrm{dA} + \int_{\mathcal{T}_i} v\, \tau \left(H\frac{\partial P}{\partial T} + P\frac{\partial H}{\partial T} \right)\, \mathrm{dA}$$
$$- \int_{\partial \mathcal{T}_i \backslash \partial \mathcal{S}} v\, (Q\, PH^3\, \nabla P) \cdot \mathbf{n}\, \mathrm{dS} = 0 \quad (5.5)$$

This is the elemental weak form of the dynamic generalized Reynolds Equation.

5.3.3   Discretization in time

For the temporal discretization of the dynamic generalized Reynolds Equation, we use the trapezoidal rule [see Hairer and Wanner, 1996]. This gives:

$$\dot{P}_{n+1} = 2 \frac{P_{n+1} - P_n}{\Delta T_{n+1}} - \dot{P}_n \qquad (5.6)$$

where $\Delta T_{n+1} = T_{n+1} - T_n$. Substituting this in Eqn. 5.5, we get:

$$\int_{T_i} \nabla v_{n+1} \cdot (Q_{n+1}\, P_{n+1} H_{n+1}^3\, \nabla P_{n+1})\; \mathrm{dA} + \int_{T_i} v_{n+1}\, \mathbf{\Lambda}_{n+1} \cdot \nabla (P_{n+1} H_{n+1})\; \mathrm{dA}$$

$$+ \int_{T_i} v_{n+1}\, \tau \left[H_{n+1} \left(2 \frac{P_{n+1} - P_n}{\Delta T_{n+1}} - \dot{P}_n \right) + P_{n+1} \dot{H}_{n+1} \right]\; \mathrm{dA}$$

$$- \int_{\partial T_i \backslash \partial S} v_{n+1}\, (Q_{n+1}\, P_{n+1} H_{n+1}^3\, \nabla P_{n+1}) \cdot \mathbf{n}\; \mathrm{dS} = 0 \quad (5.7)$$

For simplicity, we drop the subscript from all terms evaluated at T_{n+1} (thus write P_{n+1} as simply P) and add the subscript '-1' for all terms evaluated at T_n (thus P_n will be written as P_{-1}).

5.3.4   Linearization

We see that the weak form of the dynamic generalized Reynolds Equation is non-linear. To solve the nonlinear problem, we utilize the Newton Raphson scheme [see Iserles, 1996], whereby we iteratively solve a series of linearized problems to obtain the solution to the nonlinear problem. In this section we will linearize the weak form presented in Equations 5.5. Consider the linearization of Eqn. 5.5 with respect to the pressure P about the pressure P_0[1]. Writing $P = P_0 + \partial P$ and retaining only the linear terms in ∂P we get:

$$\int_{T_i} \nabla v \cdot \left(Q\ P_0 H^3\ \nabla P_0 + Q\ P_0 H^3\ \nabla \partial P + Q\ \partial P H^3\ \nabla P_0\right)\ \mathrm{dA}$$

$$+ \int_{T_i} v\ \mathbf{\Lambda} \cdot \left(H\ \nabla P_0 + P_0\ \nabla H + H\ \nabla \partial P + \partial P\ \nabla H\right)\ \mathrm{dA}$$

$$+ \int_{T_i} v\ \tau \left(2H\ \frac{P_0 - P_{-1}}{\Delta T} - H\dot{P}_{-1} + 2H\frac{\partial P}{\Delta T} + P_0\ \dot{H} + \partial P\ \dot{H}\right)\ \mathrm{dA}$$

$$- \int_{\partial T_i \backslash \partial S} v\ \left(Q\ P_0 H^3\ \nabla P_0 + Q\ P_0 H^3\ \nabla \partial P + Q\ \partial P\ H^3\ \nabla P_0\right) \cdot \mathbf{n}\ \mathrm{dS} = 0 \quad (5.8)$$

We will use Equation 5.8 to solve the nonlinear forms using the Newton Raphson scheme.

5.3.5 Streamline Upwind/Petrov-Galerkin formulation

As seen for the static case in Chapter 4, spurious oscillations are also observed when a regular non-stabilized finite element formulations is used to solve the transient advection-diffusion equation. Hence the SUPG stabilization technique proposed by Brooks and Hughes [1982] discussed in the previous section is also implemented in the dynamic case.

5.3.6 Finite element discretization

As discussed for the steady state case, in order to solve for the pressure field numerically, we approximate it to have a piecewise linear form over each triangular element. We eventually obtain $P = P_i \phi_i^e$ for $i = 1, 2, 3$, where ϕ_i^e are the local basis functions corresponding to node i. Similarly for the test functions, locally over each element, again

---

[1] P_0 is distinct from atmospheric non-dimensional pressure

we have $v = v_i \widetilde{\phi}_i^e$ for $i = 1, 2, 3$ where $\widetilde{\phi}_i^e$ are the local test basis functions corresponding to node i. Substituting these in Eqn. 5.8, we get:

$$
\int_{T_i} \left(\overline{\mathbf{B}}^{eT} \mathbf{v}^e \right)^T (Q \, P_0 H^3 \, \nabla P_0 + Q \, P_0 H^3 \, \mathbf{B}^{eT} \partial \mathbf{P}^e + Q \, \phi^{eT} \partial \mathbf{P}^e H^3 \, \nabla P_0) \, \mathrm{dA}
$$

$$
+ \int_{T_i} \widetilde{\phi}^{eT} \mathbf{v}^e \, \Lambda \cdot (H \, \nabla P_0 + P_0 \, \nabla H + H \, \mathbf{B}^{eT} \partial \mathbf{P}^e + \phi^{eT} \partial \mathbf{P}^e \, \nabla H) \, \mathrm{dA}
$$

$$
+ \int_{T_i} \widetilde{\phi}^{eT} \mathbf{v}^e \, \tau \left[\frac{2H}{\Delta T} (P_0 - P_{-1}) - H \dot{P}_{-1} + \frac{2H}{\Delta T} \phi^{eT} \partial \mathbf{P}^e + P_0 \, \dot{H} + \dot{H} \, \phi^{eT} \partial \mathbf{P}^e \right] \, \mathrm{dA}
$$

$$
- \int_{\partial T_i \backslash \partial \mathcal{S}} \widetilde{\phi}^{eT} \mathbf{v}^e \, (Q \, P_0 H^3 \, \nabla P_0 + Q \, P_0 H^3 \, \mathbf{B}^{eT} \partial \mathbf{P}^e + Q \, \phi^{eT} \partial \mathbf{P}^e H^3 \, \nabla P_0) \cdot \mathbf{n} \, \mathrm{dS} = 0 \quad (5.9)
$$

Collecting the terms we can rewrite this as:

$$
\mathbf{v}^{eT} \int_{T_i} \left[\begin{array}{c} Q P_0 H^3 \, \overline{\mathbf{B}}^e \mathbf{B}^{eT} + Q H^3 \, \overline{\mathbf{B}}^e \nabla P_0 \, \phi^{eT} + H \, \widetilde{\phi}^e \, \Lambda^T \mathbf{B}^{eT} \\[2mm] + \widetilde{\phi}^e \Lambda^T \nabla H \, \phi^{eT} + \tau \left(\dot{H} + \frac{2H}{\Delta T} \right) \widetilde{\phi}^e \phi^{eT} \end{array} \right] \mathrm{dA} \, \partial \mathbf{P}^e
$$

$$
+ \mathbf{v}^{eT} \int_{T_i} \left[\begin{array}{c} Q P_0 H^3 \, \overline{\mathbf{B}}^e \nabla P_0 + H \, \widetilde{\phi}^e \, \Lambda^T \nabla P_0 + P_0 \, \widetilde{\phi}^e \Lambda^T \nabla H \\[2mm] + \tau \widetilde{\phi}^e H \dot{P}_{-1} + \tau \widetilde{\phi}^e \frac{2H}{\Delta T} (P_0 - P_{-1}) + \tau \widetilde{\phi}^e P_0 \dot{H} \end{array} \right] \mathrm{dA}
$$

$$
- \mathbf{v}^{eT} \int_{\partial T_i \backslash \partial \mathcal{S}} \widetilde{\phi}^e \left(Q \, P_0 H^3 \, \nabla P_0 + Q \, P_0 H^3 \, \mathbf{B}^{eT} \partial \mathbf{P}^e + Q \, \phi^{eT} \partial \mathbf{P}^e H^3 \, \nabla P_0 \right) \cdot \mathbf{n} \, \mathrm{dS} = 0 \quad (5.10)
$$

Again, since the test functions are arbitrary, Eqn. 5.10 reduces to:

$$
\mathbf{K}_{dyn}^e \, \partial \mathbf{P}^e - \mathbf{R}_{dyn}^e - \int_{\partial T_i \backslash \partial \mathcal{S}} \widetilde{\phi}^e \, (Q \, P_0 H^3 \, \nabla P_0 + Q \, P_0 H^3 \, \mathbf{B}^{eT} \partial \mathbf{P}^e + Q \, \phi^{eT} \partial \mathbf{P}^e H^3 \, \nabla P_0) \cdot \mathbf{n} \, \mathrm{dS} = 0
$$

$$(5.11)$$

where $\mathbf{K}_{dyn}^e$ and $\mathbf{R}_{dyn}^e$ are the element stiffness matrix and element flux vector defined as:

$$
\mathbf{K}_{dyn}^e = \int_{T_i} \left[\begin{array}{c} Q P_0 H^3 \, \overline{\mathbf{B}}^e \mathbf{B}^{eT} + Q H^3 \, \overline{\mathbf{B}}^e \nabla P_0 \, \phi^{eT} + H \, \widetilde{\phi}^e \, \Lambda^T \mathbf{B}^{eT} \\[2mm] + \widetilde{\phi}^e \Lambda^T \nabla H \, \phi^{eT} + \tau \left(\dot{H} + \frac{2H}{\Delta T} \right) \widetilde{\phi}^e \phi^{eT} \end{array} \right] \mathrm{dA} \qquad (5.12)
$$

$$
\mathbf{R}_{dyn}^e = - \int_{T_i} \left[\begin{array}{c} Q P_0 H^3 \, \overline{\mathbf{B}}^e \nabla P_0 + H \, \widetilde{\phi}^e \, \Lambda^T \nabla P_0 + P_0 \, \widetilde{\phi}^e \Lambda^T \nabla H \\[2mm] + \tau \widetilde{\phi}^e H \dot{P}_{-1} + \tau \widetilde{\phi}^e \frac{2H}{\Delta T} (P_0 - P_{-1}) + \tau \widetilde{\phi}^e P_0 \dot{H} \end{array} \right] \mathrm{dA} \qquad (5.13)
$$

5.3.7 Mesh generation

The Reynolds equation is solved in a Lagrangian frame and the spatial mesh is not modified during the temporal solution. The spatial mesh used is obtained from the solution to the steady state problem discussed in Chapter 4.

5.3.8 Assembly

In order to obtain the complete pressure profile over the slider $\mathcal{S}$, we need to solve Eqn. 5.11 simultaneously over all of the elements. Thus the equations are assembled to form the global stiffness matrix and the global flux vector. During assembly, the flux discontinuities between the elements, accounted for by the third term in Eqn. 5.11, are neglected and hence the global system of equations is obtained as:

$$\mathbf{K} \, \partial \mathbf{P} = \mathbf{R} \tag{5.14}$$

where, $\mathbf{K} = \mathbf{K}_{dyn} = \overset{N_e}{\underset{i=1}{\mathbf{A}}} \mathbf{K}^e_{dyn}$, $\mathbf{R} = \mathbf{R}_{dyn} = \overset{N_e}{\underset{i=1}{\mathbf{A}}} \mathbf{R}^e_{dyn}$ for the dynamic case, $\partial \mathbf{P} = \overset{N_e}{\underset{i=1}{\mathbf{A}}} \partial \mathbf{P}^e$

and $\mathbf{A}$ is the assembly operator.

5.3.9 Solution

The resulting sparse linear system of equations is renumbered to form a banded matrix and the resulting system is solved using a preconditioned GMRES technique similar to the one for the static case.

5.3.10   Air bearing damping/stiffness calculation

The air-bearing damping and stiffness matrices are defined as the changes in forces and moments of the air-bearing due to changes in the flying attitude and velocities of the slider. To determine these, we need to find the change in pressure P due to changes in the attitude/velocity of the slider. The clearance H under the slider depends on the attitude of the slider as:

$$H = d_{etch} + z_{pivot} + X \cdot \theta_{pitch} + Y \cdot \theta_{roll} = H(z_{pitch}, \theta_{pitch}, \theta_{roll}) \tag{5.15}$$

where d_{etch} is the etch depth, z_{pivot} is the z-height of the pivot location, θ_{pitch} is the pitch angle and θ_{roll} is the roll angle, and X and Y are the coordinates of the point measured from the pivot location (see Fig. 5.1). Similarly the time derivatives of the clearance $\dot{H}$ depend on the velocity of the slider as:

$$\dot{H} = \dot{z}_{pivot} + X \cdot \dot{\theta}_{pitch} + Y \cdot \dot{\theta}_{roll} = \dot{H}(\dot{z}_{pitch}, \dot{\theta}_{pitch}, \dot{\theta}_{roll}) \tag{5.16}$$

Now we can write the weak form of the generalized Reynolds equation, Eqn. 5.2 as some function ϑ of pressures, clearance and their derivatives as:

$$\vartheta(P, H, \dot{P}, \dot{H}) = 0 \tag{5.17}$$

To calculate the stiffness, we differentiate this expression with respect to the slider attitude. Differentiating with respect to z_{pivot}, we get:

$$\frac{d\vartheta(P, H, \dot{P}, \dot{H})}{dz_{pivot}} = \frac{\partial\vartheta(P, H, \dot{P}, \dot{H})}{\partial H} \cdot \frac{\partial H}{\partial z_{pitch}} + \frac{\partial\vartheta(P, H, \dot{P}, \dot{H})}{\partial P} \cdot \frac{dP}{dz_{pivot}} = 0 \tag{5.18}$$

$$\Rightarrow \frac{dP}{dz_{pivot}} = \left[\frac{\partial\vartheta(P, H, \dot{P}, \dot{H})}{\partial P}\right]^{-1} \left\{\frac{\partial\vartheta(P, H, \dot{P}, \dot{H})}{\partial H} \cdot \frac{\partial H}{\partial z_{pivot}}\right\} \tag{5.19}$$

In the above formulation, we are neglecting the dependence of $\dot{P}$ on the slider's attitude H. Similarly differentiating with respect to θ_{pitch} and θ_{roll}, we obtain:

$$\frac{dP}{d\theta_{pitch}} = \left[\frac{\partial \vartheta(P,H,\dot{P},\dot{H})}{\partial P}\right]^{-1} \left\{ \frac{\partial \vartheta(P,H,\dot{P},\dot{H})}{\partial H} \cdot \frac{\partial H}{\partial \theta_{pitch}} \right\} \tag{5.20}$$

$$\frac{dP}{d\theta_{roll}} = \left[\frac{\partial \vartheta(P,H,\dot{P},\dot{H})}{\partial P}\right]^{-1} \left\{ \frac{\partial \vartheta(P,H,\dot{P},\dot{H})}{\partial H} \cdot \frac{\partial H}{\partial \theta_{roll}} \right\} \tag{5.21}$$

Substituting the finite element interpolations (Eqns. 4.10, 4.11) and evaluating the expressions above, we get:

$$\left\{ \frac{d\mathbf{P}}{dz_{pivot}} \right\} = \left[\mathbf{K}_{stat} \right]^{-1} \left\{ \frac{\partial \mathbf{R}_{stat}}{\partial z_{pivot}} \right\} \tag{5.22}$$

$$\left\{ \frac{d\mathbf{P}}{d\theta_{pitch}} \right\} = \left[\mathbf{K}_{stat} \right]^{-1} \left\{ \frac{\partial \mathbf{R}_{stat}}{\partial \theta_{pitch}} \right\} \tag{5.23}$$

$$\left\{ \frac{d\mathbf{P}}{d\theta_{roll}} \right\} = \left[\mathbf{K}_{stat} \right]^{-1} \left\{ \frac{\partial \mathbf{R}_{stat}}{\partial \theta_{roll}} \right\} \tag{5.24}$$

where $\mathbf{K}_{stat}$ is the global steady state stiffness matrix and the vectors $\frac{\partial \mathbf{R}_{stat}}{\partial z_{pivot}}$, $\frac{\partial \mathbf{R}_{stat}}{\partial \theta_{pitch}}$ and $\frac{\partial \mathbf{R}_{stat}}{\partial \theta_{roll}}$ are defined as:

$$\left\{ \frac{\partial \mathbf{R}_{stat}}{\partial z_{pivot}} \right\} = - \mathop{\mathbf{A}}_{i=1}^{N_e} \int_{T_i} (3\,QP_0H^2\,\overline{\mathbf{B}}^e \nabla P_0 + \tilde{\phi}^e \mathbf{\Lambda}^T \nabla P_0)\,\mathrm{d}A \tag{5.25}$$

$$\left\{ \frac{\partial \mathbf{R}_{stat}}{\partial \theta_{pitch}} \right\} = - \mathop{\mathbf{A}}_{i=1}^{N_e} \int_{T_i} (3\,QP_0H^2\,\overline{\mathbf{B}}^e \nabla P_0 + \tilde{\phi}^e \mathbf{\Lambda}^T \nabla P_0) \cdot X\,\mathrm{d}A \tag{5.26}$$

$$\left\{ \frac{\partial \mathbf{R}_{stat}}{\partial \theta_{roll}} \right\} = - \mathop{\mathbf{A}}_{i=1}^{N_e} \int_{T_i} (3\,QP_0H^2\,\overline{\mathbf{B}}^e \nabla P_0 + \tilde{\phi}^e \mathbf{\Lambda}^T \nabla P_0) \cdot Y\,\mathrm{d}A \tag{5.27}$$

The above expressions hold for both the time dependent and the steady state versions of the Reynolds equation. The terms of the 3×3 stiffness matrix can then be evaluated as:

$$
\mathbf{K}_{abs} =
\begin{bmatrix}
\mathbf{C}_{F_z}^T \left\{ \dfrac{\partial \mathbf{R}_{stat}}{\partial z_{pivot}} \right\} & \mathbf{C}_{M_{pitch}}^T \left\{ \dfrac{\partial \mathbf{R}_{stat}}{\partial z_{pivot}} \right\} & \mathbf{C}_{M_{roll}}^T \left\{ \dfrac{\partial \mathbf{R}_{stat}}{\partial z_{pivot}} \right\} \\[3mm]
\mathbf{C}_{F_z}^T \left\{ \dfrac{\partial \mathbf{R}_{stat}}{\partial \theta_{pitch}} \right\} & \mathbf{C}_{M_{pitch}}^T \left\{ \dfrac{\partial \mathbf{R}_{stat}}{\partial \theta_{pitch}} \right\} & \mathbf{C}_{M_{pitch}}^T \left\{ \dfrac{\partial \mathbf{R}_{stat}}{\partial \theta_{pitch}} \right\} \\[3mm]
\mathbf{C}_{F_z}^T \left\{ \dfrac{\partial \mathbf{R}_{stat}}{\partial \theta_{roll}} \right\} & \mathbf{C}_{M_{pitch}}^T \left\{ \dfrac{\partial \mathbf{R}_{stat}}{\partial \theta_{roll}} \right\} & \mathbf{C}_{M_{roll}}^T \left\{ \dfrac{\partial \mathbf{R}_{stat}}{\partial \theta_{roll}} \right\}
\end{bmatrix}
\tag{5.28}
$$

Thus the stiffness is obtained by the solution of three extra linear systems. However this is not computationally very expensive even with iterative methods (where the system matrix $\mathbf{K}$ has not been factorized) since preconditioners for $\mathbf{K}$ will already have been evaluated.

Now we consider the damping matrix evaluation. Again we consider Eqn. 5.17. This time, we differentiate with respect to the time derivative of the clearance $\dot{H}$ and velocities of the slider attitudes. Differentiating with respect to $\dot{z}_{pivot}$, we obtain:

$$
\frac{d\vartheta(P, H, \dot{P}, \dot{H})}{d\dot{z}_{pivot}} = \frac{\partial\vartheta(P, H, \dot{P}, \dot{H})}{\partial \dot{H}} \cdot \frac{\partial \dot{H}}{\partial \dot{z}_{pitch}} + \frac{\partial\vartheta(P, H, \dot{P}, \dot{H})}{\partial P} \cdot \frac{dP}{d\dot{z}_{pivot}} = 0
\tag{5.29}
$$

$$
\Rightarrow \frac{dP}{d\dot{z}_{pivot}} = \left[\frac{\partial\vartheta(P, H, \dot{P}, \dot{H})}{\partial P} \right]^{-1} \left\{ \frac{\partial\vartheta(P, H, \dot{P}, \dot{H})}{\partial \dot{H}} \cdot \frac{\partial \dot{H}}{\partial \dot{z}_{pivot}} \right\}
\tag{5.30}
$$

Again, we are neglecting the dependence of $\dot{P}$ on $\dot{H}$. This assumption is reasonable when the slider is close to the steady state fly height. Similarly differentiating with respect to $\dot{\theta}_{pitch}$ and $\dot{\theta}_{roll}$, we obtain:

$$
\frac{dP}{d\dot{\theta}_{pitch}} = \left[\frac{\partial\vartheta(P, H, \dot{P}, \dot{H})}{\partial P} \right]^{-1} \left\{ \frac{\partial\vartheta(P, H, \dot{P}, \dot{H})}{\partial \dot{H}} \cdot \frac{\partial \dot{H}}{\partial \dot{\theta}_{pitch}} \right\}
\tag{5.31}
$$

$$
\frac{dP}{d\dot{\theta}_{roll}} = \left[\frac{\partial\vartheta(P, H, \dot{P}, \dot{H})}{\partial P} \right]^{-1} \left\{ \frac{\partial\vartheta(P, H, \dot{P}, \dot{H})}{\partial \dot{H}} \cdot \frac{\partial \dot{H}}{\partial \dot{\theta}_{roll}} \right\}
\tag{5.32}
$$

Substituting the finite element interpolations (Eqns. 4.10, 4.11) and evaluating the expressions above, we get:

$$\left\{ \frac{d\mathbf{P}}{d\dot{z}_{pivot}} \right\} = [\mathbf{K}_{stat}]^{-1} \left\{ \frac{\partial \mathbf{R_{dyn}}}{\partial \dot{z}_{pivot}} \right\} \tag{5.33}$$

$$\left\{ \frac{d\mathbf{P}}{d\dot{\theta}_{pitch}} \right\} = [\mathbf{K}_{stat}]^{-1} \left\{ \frac{\partial \mathbf{R_{dyn}}}{\partial \dot{\theta}_{pitch}} \right\} \tag{5.34}$$

$$\left\{ \frac{d\mathbf{P}}{d\dot{\theta}_{roll}} \right\} = [\mathbf{K}_{stat}]^{-1} \left\{ \frac{\partial \mathbf{R_{dyn}}}{\partial \dot{\theta}_{roll}} \right\} \tag{5.35}$$

where $\mathbf{K}_{stat}$ is the global steady state stiffness matrix and the vectors $\frac{\partial \mathbf{R}_{dyn}}{\partial \dot{z}_{pivot}}$, $\frac{\partial \mathbf{R}_{dyn}}{\partial \dot{\theta}_{pitch}}$ and $\frac{\partial \mathbf{R}_{dyn}}{\partial \dot{\theta}_{roll}}$ are defined as:

$$\left\{ \frac{\partial \mathbf{R_{dyn}}}{\partial \dot{z}_{pivot}} \right\} = - \mathop{\mathbf{A}}_{i=1}^{N_e} \int_{T_i} (\tau P \widetilde{\phi}^e) \, \mathrm{dA} \tag{5.36}$$

$$\left\{ \frac{\partial \mathbf{R_{dyn}}}{\partial \dot{\theta}_{pitch}} \right\} = - \mathop{\mathbf{A}}_{i=1}^{N_e} \int_{T_i} (\tau P \widetilde{\phi}^e) \cdot X \, \mathrm{dA} \tag{5.37}$$

$$\left\{ \frac{\partial \mathbf{R_{dyn}}}{\partial \dot{\theta}_{roll}} \right\} = - \mathop{\mathbf{A}}_{i=1}^{N_e} \int_{T_i} (\tau P \widetilde{\phi}^e) \cdot Y \, \mathrm{dA} \tag{5.38}$$

The terms of the 3×3 damping matrix can then be evaluated as:

$$\mathbf{C}_{abs} = \begin{bmatrix} \mathbf{C}_{F_z}^T \left\{ \frac{\partial \mathbf{R}_{dyn}}{\partial \dot{z}_{pivot}} \right\} & \mathbf{C}_{M_{pitch}}^T \left\{ \frac{\partial \mathbf{R}_{dyn}}{\partial \dot{z}_{pivot}} \right\} & \mathbf{C}_{M_{roll}}^T \left\{ \frac{\partial \mathbf{R}_{dyn}}{\partial \dot{z}_{pivot}} \right\} \\ \mathbf{C}_{F_z}^T \left\{ \frac{\partial \mathbf{R}_{dyn}}{\partial \dot{\theta}_{pitch}} \right\} & \mathbf{C}_{M_{pitch}}^T \left\{ \frac{\partial \mathbf{R}_{dyn}}{\partial \dot{\theta}_{pitch}} \right\} & \mathbf{C}_{M_{pitch}}^T \left\{ \frac{\partial \mathbf{R}_{dyn}}{\partial \dot{\theta}_{pitch}} \right\} \\ \mathbf{C}_{F_z}^T \left\{ \frac{\partial \mathbf{R}_{dyn}}{\partial \dot{\theta}_{roll}} \right\} & \mathbf{C}_{M_{pitch}}^T \left\{ \frac{\partial \mathbf{R}_{dyn}}{\partial \dot{\theta}_{roll}} \right\} & \mathbf{C}_{M_{roll}}^T \left\{ \frac{\partial \mathbf{R}_{dyn}}{\partial \dot{\theta}_{roll}} \right\} \end{bmatrix} \tag{5.39}$$

Thus the damping is obtained by the solution of three additional systems.

5.3.11 Algorithmic air bearing stiffness

The algorithmic stiffness matrix is defined as the change in forces and moments of the air-bearing due to changes in the flying attitude of the slider after time discretization has been done. Proceeding in a fashion similar to the previous section, we get:

$$\left\{ \frac{d\mathbf{P}}{dz_{pivot}} \right\} = [\mathbf{K}_{dyn}]^{-1} \left\{ \frac{\partial \mathbf{R}_{dyn}}{\partial z_{pivot}} \right\} \tag{5.40}$$

$$\left\{ \frac{d\mathbf{P}}{d\theta_{pitch}} \right\} = [\mathbf{K}_{dyn}]^{-1} \left\{ \frac{\partial \mathbf{R}_{dyn}}{\partial \theta_{pitch}} \right\} \tag{5.41}$$

$$\left\{ \frac{d\mathbf{P}}{d\theta_{roll}} \right\} = [\mathbf{K}_{dyn}]^{-1} \left\{ \frac{\partial \mathbf{R}_{dyn}}{\partial \theta_{roll}} \right\} \tag{5.42}$$

where $\mathbf{K}_{dyn}$ is the global dynamic stiffness matrix and the vectors $\frac{\partial \mathbf{R}_{dyn}}{\partial z_{pivot}}$, $\frac{\partial \mathbf{R}_{dyn}}{\partial \theta_{pitch}}$ and $\frac{\partial \mathbf{R}_{dyn}}{\partial \theta_{roll}}$ are defined as:

$$\left\{ \frac{\partial \mathbf{R}_{dyn}}{\partial z_{pivot}} \right\} = - \mathop{\mathbf{A}}_{i=1}^{N_e} \int_{T_i} \begin{bmatrix} 3\,QP_0H^2\,\overline{\mathbf{B}}^e\nabla P_0 + \widetilde{\phi}^e\mathbf{\Lambda}^T\nabla P_0 + P_0\,\widetilde{\phi}^e\mathbf{\Lambda}^T\mathbf{B}^{eT} \\ +\tau\widetilde{\phi}^e\dot{P}_{-1} + \tau\widetilde{\phi}^e\frac{2}{\Delta T}(2P_0 - P_{-1}) \end{bmatrix} dA \tag{5.43}$$

$$\left\{ \frac{\partial \mathbf{R}_{dyn}}{\partial \theta_{pitch}} \right\} = - \mathop{\mathbf{A}}_{i=1}^{N_e} \int_{T_i} \begin{bmatrix} 3\,QP_0H^2\,\overline{\mathbf{B}}^e\nabla P_0 + \widetilde{\phi}^e\mathbf{\Lambda}^T\nabla P_0 + P_0\,\widetilde{\phi}^e\mathbf{\Lambda}^T\mathbf{B}^{eT} \\ +\tau\widetilde{\phi}^e\dot{P}_{-1} + \tau\widetilde{\phi}^e\frac{2}{\Delta T}(2P_0 - P_{-1}) \end{bmatrix} \cdot X\,dA \tag{5.44}$$

$$\left\{ \frac{\partial \mathbf{R}_{dyn}}{\partial \theta_{roll}} \right\} = - \mathop{\mathbf{A}}_{i=1}^{N_e} \int_{T_i} \begin{bmatrix} 3\,QP_0H^2\,\overline{\mathbf{B}}^e\nabla P_0 + \widetilde{\phi}^e\mathbf{\Lambda}^T\nabla P_0 + P_0\,\widetilde{\phi}^e\mathbf{\Lambda}^T\mathbf{B}^{eT} \\ +\tau\widetilde{\phi}^e\dot{P}_{-1} + \tau\widetilde{\phi}^e\frac{2}{\Delta T}(2P_0 - P_{-1}) \end{bmatrix} \cdot Y\,dA \tag{5.45}$$

Again, the terms of the 3×3 stiffness matrix can then be evaluated as:

$$\mathbf{K}_{alg} = \begin{bmatrix} \mathbf{C}_{F_z}^T \left\{ \frac{\partial \mathbf{R}_{dyn}}{\partial z_{pivot}} \right\} & \mathbf{C}_{M_{pitch}}^T \left\{ \frac{\partial \mathbf{R}_{dyn}}{\partial z_{pivot}} \right\} & \mathbf{C}_{M_{roll}}^T \left\{ \frac{\partial \mathbf{R}_{dyn}}{\partial z_{pivot}} \right\} \\ \mathbf{C}_{F_z}^T \left\{ \frac{\partial \mathbf{R}_{dyn}}{\partial \theta_{pitch}} \right\} & \mathbf{C}_{M_{pitch}}^T \left\{ \frac{\partial \mathbf{R}_{dyn}}{\partial \theta_{pitch}} \right\} & \mathbf{C}_{M_{pitch}}^T \left\{ \frac{\partial \mathbf{R}_{dyn}}{\partial \theta_{pitch}} \right\} \\ \mathbf{C}_{F_z}^T \left\{ \frac{\partial \mathbf{R}_{dyn}}{\partial \theta_{roll}} \right\} & \mathbf{C}_{M_{pitch}}^T \left\{ \frac{\partial \mathbf{R}_{dyn}}{\partial \theta_{roll}} \right\} & \mathbf{C}_{M_{roll}}^T \left\{ \frac{\partial \mathbf{R}_{dyn}}{\partial \theta_{roll}} \right\} \end{bmatrix} \tag{5.46}$$

The algorithmic stiffness matrix is used in place of the stiffness and damping matrices for Newton's iterations for the equations of motion, since it is a more accurate representation of the stiffness of the time-discretized generalized Reynolds equation.

5.4 Numerical Simulations

In this section we simulate the free vibrations of the suspension/air-bearing/disk system. The initial conditions provided to the system are perturbed from the steady state and the resulting free vibrations of the system are simulated. Forced responses to events such as shock and load/unload are presented in Chapters 6 and 7.

The slider design used in this simulation is shown in Fig. 5.2. The slider fly-height, pitch and roll for the simulation are plotted in Fig. 5.3 along with z-force and pitch/roll moments in Fig. 5.4. In Fig. 5.5, we plot the error estimate and the time-step size.

5.5 Discussion

In this chapter we extended the finite element formulation developed for the steady state case in Chapter 4 to the time dependent case. We use a variable time-step discretization scheme to advance the equation in time. Finally the method is implemented and we present results for a free vibration simulation carried out using the method.

5.6   Figures

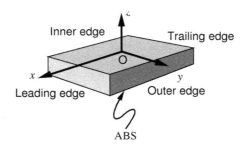

Figure 5.1: System Coordinate System

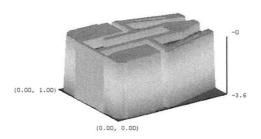

Figure 5.2: Slider 2

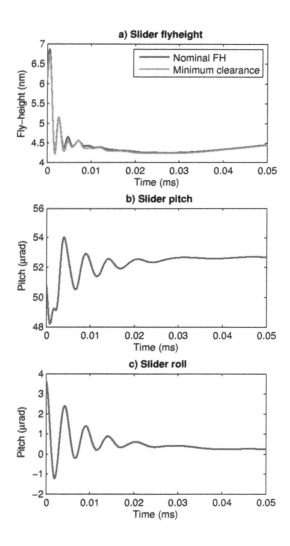

Figure 5.3: Slider attitude for free vibration simulation

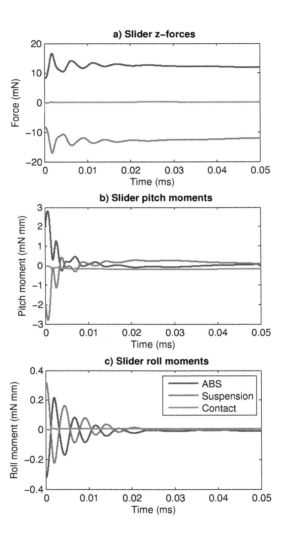

Figure 5.4: Slider forces for free vibration simulation

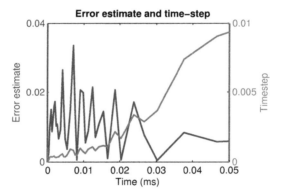

Figure 5.5: Error-estimate/time-step for free vibration simulation

Chapter 6

Simulation of the shock event

6.1 Introduction

The increased usage of small form factor hard disk drives in hostile environments such as MP3 players, cameras etc. as well as the need to improve operating-shock specifications for traditional applications such as laptops have led to a need to design better head/suspension systems. In order to aid in the design process, there is a need to quickly and accurately simulate the operating shock response of the hard disk drive system. In this chapter we present a method for simulating shock for the suspension/air-bearing/disk system implemented on the dynamic formulation discussed in Chapters 2 and 5. We also present results from various parametric studies carried out using a previous implementation using finite volume modeling [developed by Hu and Bogy, 1995] for the air-bearing, but using the structural method discussed in Chapter 2 developed by the author.

Zeng and Bogy [2000c] mentioned that there are essentially three approaches for dealing with shock problems. The first being the design and installation of a suitable isolation system for the disk drives. The second, to design a robust servo control mechanism to prevent read/write errors during shock and the third is to design a robust mechanical system such that the head-disk interface is resistant to shock. A combination of these three techniques needs to be used to effectively counter the problem of shock.

6.2 Prior Work

Over the past few years there have been various experimental and simulation studies on the shock response of the mechanical system and its effects on the head-disk interface. Many of these studies [Harrison and Mundt, 2000; Edwards, 1999; Kumar, Khanna, and Sri-Jayantha, 1994; Kouhei, Yamada, Keroba, and Aruga, 1995] have been limited to the non-operating state of the drives, and/or to the component level. Various other papers [Jayson, Murphy, Smith, and Talke, 2003; Jiang, Takashima, and Chonan, 1995] have considered shock simulations in the operating state using simplified models for one or more components of the drive, i.e. either the disk,suspension or the air bearing. A summary of these studies has been presented by Bhargava and Bogy [2007b].

For the simulation of operational shock Zeng and Bogy [2000c] proposed a method whereby they separate the simulation work into two essentially uncoupled sets. They developed a finite element model of the disk and suspension system and used it to obtain the dynamic normal load and moments applied to the slider air bearing. These were then used as input data for an air bearing dynamic simulator to calculate the dynamic flying attitudes. They were able to obtain not only the responses of the structural components, but also the responses of the slider air bearings. Also, simulations where the air bearing exhibits highly nonlinear behavior, such as when the air bearing collapses, may require iterations between the structural and air bearing simulations, thereby making the process cumbersome and computationally more expensive. In a previous paper, Bhargava and Bogy [2007b] proposed a method in which the modeling of the structural components was performed in ANSYS, a commercial finite element package. The air bearing modeling used the CML dynamic air bearing simulator. The two modules are coupled and each is iterated to convergence at every time step. The pulse width of the shock was kept constant at 0.5 ms and the magnitude of the shock pulse was varied. However, this method was inefficient and computationally expensive due to the exchange of data between the two modules at each time step.

Here we propose an improved simulation method whereby the structural modeling module is transferred into the air bearing code using pre-assembled exported mass and stiffness matrices from ANSYS. This method is found to be as accurate, though much faster and more robust, than the one proposed earlier. Using this new approach, we simulate the effect of the pulse width on the shock resistance of a 1" drive.

6.3 The CML L/UL/S Simulator

This study was carried out using the CML Load/Unload/Shock simulator also developed by the author. The simulator was released for use by CML members in July, 2006. The manual for the simulator is attached in Appendix A. A brief outline of the procedure used by this simulator is described in this section.

6.3.1 Suspension modeling

The suspension modeling is carried out in the same manner as described in Chapter 2. A finite element model of the suspension is used to generate reduced stiffness and mass matrices. The nonlinear contact constraints are imposed on top of the stiffness matrix using the technique described in Hughes et al. [1976].

6.3.2 Air bearing modeling

The air bearing modeling is done differently from the procedure described in Chapter 5. The air bearing modeling is based on Patankar's [Patankar, 1980] finite volume method. This procedure has been documented in detail in several CML reports [Hu and Bogy, 1995]. However, in addition, 6 degrees of freedom (DOF) are used here rather than just three to model the slider state: the displacements and rotations in each of the 3 directions (x, y and z).

6.3.3 Coupling

The structural and air bearing components are coupled by a fixed point iteration scheme that iterates all forces and displacements at each time step to convergence. The basic scheme is shown in Fig. 6.12.

6.4   Methodology

6.4.1   Modeling Shock

A hard disk drive experiences shock when it is suddenly brought to rest from a state of motion, such as during a drop on to a hard surface. This sudden deceleration of the system is modeled as a body force field pulse to the structural components of the system. Numerically this pulse is approximated to have the form of a half sine wave (see 6.1). In this study we investigate shocks in the z-direction only, i.e. shocks in a direction perpendicular to the spinning disk. In the z-direction, we can have two types of shocks: positive and negative (where the acceleration pulse is applied in the $+z$ and $-z$ directions respectively). In a previous study, Bhargava and Bogy [2007b] showed that drives are less robust for positive shocks than for negative shocks. Hence in this chapter we discuss only positive shocks.

6.5   Pulse-width effect

In this section we study the mechanism of failure of the head-disk interface during shock. The simulation program developed earlier in this chapter is used to simulate the shock response of a system for various different kinds of shock, which are characterized by varying pulse widths for their acceleration pulse. This allows us to simulate drops on surfaces of varying stiffness. We simulate shocks of pulse widths in the range of 0.2 ms to 3.0 ms. A shock of 0.2 ms could correspond to the disk-drive falling on a concrete pavement (depending on the amount of shock isolation provided in the drive) and 3.0 ms could correspond to the drive falling on a carpeted floor. As we increase the pulse width of the shock the response tends to become more quasi static in nature, with the acceleration pulse leading simply to a gram load change for the air bearing, but with little dynamic effects. Hence we limit the study to a maximum pulse width of 3.0 ms.

In the following subsections we start with a discussion on the dynamics of the individual structural components, i.e. the disk and the suspension, followed by a presentation of our simulation results and an explanation of those results based on the dynamics of the individual components, followed by conclusions and remarks at the end of the section.

6.5.1 Component Dynamics

In this section we discuss the dynamics and the shock response of the main structural components of the disk drive, i.e. the disk and the suspension.

Disk

To study the effect of the shock pulse width on the response of the disk drive, we first study the response of the disk to shocks of varying pulse widths (see Fig. 6.1 for a description of the shock pulse). The parameters of the disk are given in Table 6.1. It has been shown in various studies [Zeng and Bogy, 2000c; Bhargava and Bogy, 2007b] that the shock response of a rotating disk to an axisymmetric shock is primarily composed of the first axisymmetric (umbrella) mode (Fig. 6.2). The only effect of the rotation of the disk on the axisymmetric modes is due to centrifugal stiffening. For low speeds of rotation, such as 3600 RPM, and small diameters of the disk, this effect is negligible and hence the disk can be modeled as stationary for these cases. The z-displacement of a point on the OD of the disk, when subject to an acceleration pulse of 200G magnitude and varying pulse width is plotted in Fig. 6.3. We observe that even though the magnitude of the acceleration impulse is 200G in each case, the response of the disk is very different for different pulse widths. We observe that for a short pulse width like 0.2 ms, much more energy is transferred to the disk than for a pulse width of 0.5 ms. The maximum amplitudes during the shock and during the post-shock response of the disk are plotted in Fig. 6.4. We see that the deflection in the post-shock response of the disk is a strong function of the pulse width of the shock. For example, with a pulse width of 0.5 ms, even though the disk might deflect more than 6 μm during the shock, the disk only oscillates at less than 0.5 μm after the shock. However for a pulse width of 0.2 ms the disk deflection during and after the shock both are large. We will see that such large disk oscillations in the post-shock stage can lead to failure of the head-disk interface due to resonance with the suspension. Fig. 6.5 shows a waterfall plot of the frequency spectra of the disk response to a 200G shock of varying pulse widths. We see that the primary mode excited in all cases, is the first umbrella mode (see Fig. 6.2) at 3.04 KHz. However, the amount of power in this mode varies strongly with the pulse width, as also seen in Fig. 6.4.

Suspension

The suspension is one of the most important structural components in the hard disk drive. A schematic diagram of the suspension is shown in Fig. 6.6. In a 1" drive, the suspension attaches directly to the actuator hub.

For the simulations we used a suspension model from a popular 1" drive. The first three suspension modes of vibration, the first and second bending modes and the first torsion mode, are plotted in Figs. 6.7-6.9. We see that the frequency of the second bending mode, shown in Fig. 6.9, is 3.12 KHz, which is very close to the umbrella mode of vibration of the disk. In Fig. 6.10 we plot the free suspension shock response (displacement of slider center) to a 200G shock of varying pulse widths. In Fig. 6.11 we plot the power spectra of these responses. We see that the primary mode of vibration for all pulse widths is the first bending mode, with a frequency of 320.6 Hz. For a pulse width of 0.2 ms there is also some power in the second bending mode, which is at around 3.12 KHz. However, we will see that when the suspension is loaded onto the disk, the second bending mode can get excited due to resonance with the umbrella mode of the disk, which can lead to failure of the head-disk interface.

6.5.2   Results and Discussion

Various simulations were carried out to investigate the effect of the pulse width on the shock response of the system. In the simulations the slider design is a 'femco' slider from a popular 1" disk drive. The design is shown in Fig. 6.13 and its operating parameters are listed in Table 6.2.

Fig. 6.14 shows the slider attitude response to a 125G shock of 0.2 ms pulse width. The various quantities plotted are, the shock profile in a), the absolute displacement of the slider center, the load/unload (L/UL) tab and the disk in b), the nominal fly height in c), the minimum clearance in d) and the roll and the pitch in e) and f) respectively. In b), we see that the slider center follows the displacement of the disk perfectly. However for the L/UL

tab, we see that its oscillations (which correspond to the oscillations of the load beam) grow

in time, which is due to the resonance between the disk and the suspension. In Fig. 6.15,

we see that the frequency of the oscillations of the load beam as well as the disk are close to

3 KHz, which corresponds to the umbrella mode of the disk and the second bending mode

of the suspension. In Fig. 6.14 d), we see that this resonance eventually causes head disk

contact. In Fig. 6.16 we plot the contact forces and separations of the dimple and the two

limiters (referred to as CE1 and CE2). We can see that as the oscillations of the load beam

increase, the dimple begins to open and close repeatedly. Eventually the impacts between

the load beam and the flexure become strong enough to cause the slider to contact the disk.

The forces corresponding to this response are plotted in Fig. 6.17. In a), we plot the air

bearing forces, positive, negative and total and in b) we plot the asperity contact forces

(quasi-static, calculated when the fly height is less than the glide height) and the dynamic

impact forces (calculated when the fly height is less than zero). We see even though there is

no impact between the head and the disk during the shock, the resonance in the post-shock

response causes severe head disk impact.

In Fig. 6.18, we plot the slider attitude response to a 250G shock of 0.5 ms pulse

width. In this case, we see that the oscillations of the load beam (L/UL tab) do not grow in

time, since the disk oscillations are weak and insufficient to resonate the suspension. Hence

there is no head-disk contact in this case.

In Fig. 6.19, we plot the slider attitude response to a 375G shock of 1.0 ms pulse

width. We observe that the disk oscillations are strong enough to excite the suspension, however not strong enough to overcome damping and cause head disk contact. Hence we see that eventually the oscillations of both, the disk and the suspension die out.

If we increase the shock to 400G, we see that there is head-disk contact, not due to resonance, but because of the shock itself. This is plotted in Fig. 6.20. We see that at about 1.1 ms the load beam which springs back after being pulled by the shock, hits the flexure and causes the slider to hit the disk. This can be seen from the contact element forces and spacings in Fig. 6.21. We see that just before the slider crashes into the disk, the dimple closes with a large spike in the contact force, which signifies impact.

In Fig. 6.22 and Fig. 6.23, we plot the slider response to shocks of 400G, 2.0 ms and 450G, 3.0 ms, respectively. We see that the head-disk interface is able to withstand larger values of shock at increasing pulse widths. The reason for this is that larger pulse widths lead to lower force gradients and weaker post-shock disk response.

6.5.3 Conclusions

In this section we found that the disk response is critical to the shock resistance of the disk drive. We also found that matching suspension and disk frequencies can lead to resonance and hence failure of the head disk interface. The results of the study are summarized in Fig. 6.24. We plot the 'Safe' shock levels, i.e. the amount of shock the disk drive is able to withstand without head-disk contact as a function of the shock pulse width. We see that for short pulse widths, small magnitudes of shock are sufficient to cause head disk contacts, while for larger pulse widths, the head disk interface is able to withstand much higher amplitude shocks. However, we can avoid short pulse widths during shocks easily by providing a minimal amount of padding/isolation in a space/weight constrained system to increase the shock resistance dramatically even for shocks on the hardest surfaces. We can also avoid resonances between the suspension and the disk by properly designing the suspension such that none of the bending/torsional frequencies are close to the disk umbrella mode frequency.

6.6 Effect of dimple location

In this section we study the effect of varying the location of the dimple (pivot) of the suspension on the shock performance of a small form factor drive. Previously we have developed a shock simulator to accurately simulate the shock event and predict the response of the suspension-slider-disk system [Bhargava and Bogy, 2007c]. Here this simulator is used to simulate the shock response of a system for a suspension system with varying dimple locations in the x and y directions (see Fig.6.25). Simulations are carried out to determine the 'safe' operating shock levels, wherein no head-disk contacts occur.

6.6.1 Procedure

Fig. 6.6 shows a schematic diagram of a suspension. A typical suspension consists of a slider mounted on a flexure. The flexure provides roll and pitch stiffness to the slider, while the stiffness in the z direction is provided by the load beam through a dimple which pushes down into the flexure that adheres to the slider. Limiters may be provided on the suspension to limit the maximum separation of the flexure from the dimple during unloading. Figure 6.26 shows a free body diagram of the slider in the steady state flying condition. For simplicity, we consider here the force and moment balances in the $z - x$ plane only. Forces and moments in the third direction can be similarly balanced. We also assume that the gram load on the slider is completely applied by the dimple and moments are applied only by the flexure, thereby neglecting the z-stiffness of the flexure and the friction moments applied by the dimple. Writing the force balance in the z-direction and the moments in the y-direction, we have,

$$F_b = F_g \tag{6.1}$$

$$M_f - F_g \times x_l = M_b \tag{6.2}$$

Thus to get the same torques in the steady state (and hence the same flying attitude for the slider) for different load points x_l, the $F_g x_l$ term needs to be balanced by the M_f term, which is achieved by adjusting the pitch static attitude (PSA) of the slider to generate the additional moment. If x_l is positive, a negative PSA is used, and vice versa. Similarly in the $z - y$ plane, the roll static attitude (RSA) is adjusted to maintain the same flying roll attitude.

6.6.2   Results and Discussion

We carried out simulations for various dimple locations on the flexure for half sinusoid shocks of varying magnitudes. The operating parameters of the slider used are listed in Table 6.2.

Fig. 6.27 presents several graphs of the shock response to a 0.5 ms pulse width for the dimple located at the center of the slider (BC) ($x_l = 0.0, y_l = 0.0$) mm). In Fig. 6.28, we plot the dimple separation and dimple contact forces. It is observed that during the course of the shock, the load beam is pulled up and the dimple separates from the flexure. Subsequently, the load beam slaps back onto the flexure, hitting it with a large contact force. This causes the slider to crash into the disk. This kind of response is typical for shock pulse widths of 0.5 ms. The mechanism of failure has been previously shown to be typical of pulse widths of 0.5 - 3.0 ms [Bhargava and Bogy, 2007c].

Fig. 6.29 shows the shock response for the dimple location U1 ($x_l = -0.1$,$y_l = 0.0$ mm). We observe no head disk contact in this case. In Fig. 6.30 we plot the dimple separation, as well as the dimple contact force. As the dimple unloads, we observe in Fig. 6.29 that the pitch angle is larger than was observed for the base case (BC) in Fig. 6.27. This is a consequence of the positive PSA used to account for the dimple load moment $F_g x_l$. We also note that the minimum fly height is less than what was observed for the base case, again as a consequence of the larger pitch angle. As the load beam snaps back onto the flexure, there is a spike in the dimple contact force. This large contact force also generates a contact moment about the center of the slider.

We can explain this difference in the behavior of the two dimple location cases using a simple linearized dynamic model.

Here, z is the fly height of the center of the slider, α is the pitch and β is the roll of the slider. We linearize the air bearing for the base case at time t, with attitude $\mathbf{u_0}$ which is the instant when the load beam impacts the flexure (see Fig. 6.31). The incremental force vector F defined as the vector of restoring forces and moments generated due to changes in the attitude, is given by:

$$\mathbf{F} = \mathbf{K_b u},\qquad\qquad (6.3)$$

where the attitude vector **u** and force vector **F** are defined as:

$$\mathbf{u} = \left\{ \begin{array}{c} z \\ \alpha \\ \beta \end{array} \right\} ; \mathbf{F} = \left\{ \begin{array}{c} F_z \\ M_\alpha \\ M_\beta \end{array} \right\} \tag{6.4}$$

The matrix $\mathbf{K_b}$ is the linear spring stiffness of the air bearing, given as:

$$\mathbf{K_b} = \left\{ \begin{array}{ccc} k_{zz} & k_{z\alpha} & k_{z\beta} \\ k_{\alpha z} & k_{\alpha\alpha} & k_{\alpha\beta} \\ k_{\beta z} & k_{\beta\alpha} & k_{\beta\beta} \end{array} \right\} \tag{6.5}$$

Inverting the stiffness matrix, we obtain the compliance matrix, which relates the attitude of the slider to applied forces and moments.

$$\mathbf{C} = \mathbf{K_b}^{-1} \tag{6.6}$$

$$\mathbf{u} = \mathbf{CF} \tag{6.7}$$

We observe that during head-disk contact, the pitch is positive, while the roll can be positive or negative depending on the value of y_l, slider design, skew and other operating parameters. Thus the location of the point on the slider with the minimum fly height will be near the trailing edge. For simplicity, we assume that the minimum fly height (z_{min}) location is in fact one of the corners of the trailing edge (depending on the roll of the slider). Then we calculate the ratio of the displacement of the minimum fly height to an applied force at (x_l, y_l), i.e. $c_{z_{min}G}$. A force F_g acting at (x_l, y_l) is equivalent to the following force vector in our defined attitude coordinate system:

$$
\mathbf{F} = \left\{ \begin{array}{c} F_g \\[6pt] F_g x_l \\[6pt] F_g y_l \end{array} \right\}
\tag{6.8}
$$

Also we can calculate z_{min}, given the slider attitude vector as:

$$
z_{min} = \left\{ \begin{array}{ccc} 1 & \frac{l}{2} & sign(\beta)\frac{b}{2} \end{array} \right\} \left\{ \begin{array}{c} z \\[6pt] \alpha \\[6pt] \beta \end{array} \right\}
\tag{6.9}
$$

Here l and b are the length and breadth of the slider, and the pitch and roll are assumed to be small. Thus we have for $c_{z_{min}G}$:

$$
c_{z_{min}G} = \left\{ \begin{array}{ccc} 1 & \frac{l}{2} & \frac{b}{2} \end{array} \right\} \mathbf{C} \left\{ \begin{array}{c} F_g \\[6pt] F_g x_l \\[6pt] F_g y_l \end{array} \right\}
\tag{6.10}
$$

The $c_{z_{min}G}$ values for various values of x_l and y_l are plotted in Fig. 6.32 using K_b determined from simulations for our system. The black line on the plot is the zero displacement line, which means that the minimum fly height will not change when a small force is applied in the z direction anywhere along this line. This concept is similar to a 'center of percussion', the only difference being that we are looking at the displacement of a point other than where the force is applied. A positive value of $c_{z_{min}G}$ implies that the change in the minimum fly height will be positive for a given positive F_g applied at (x_l, y_l), while a negative value indicates that the minimum fly height will actually increase when a force F_g is applied at (x_l, y_l). We see that $c_{z_{min}G}$ is smallest when $x_l = \frac{l}{2}$ and $y_l = \pm\frac{b}{2}$.

However, earlier we noted that the pitch for the U1 case was larger than, and the minimum fly height was lower than, that of the BC, before the load beam had snapped back onto the flexure. If we choose a large negative value of x_l, the slider crashes into the disk even before the load beam snaps back onto the flexure. In Fig. 6.33 we plot the flying attitude for location U2 ($x_l = -0.2, y_l = 0.0$) mm. We observe that the mechanism of failure for this dimple location is completely different from that in the first two cases. In Fig. 6.34 we plot the dimple separation and contact force. We see that the slider crashes soon after the dimple separates. This is due to the large positive torque generated as the dimple unloads.

Various other simulations were carried out at other values of (x_l, y_l) to determine the 'safe' shock map shown in Fig. 6.35 for this particular slider-suspension system.

6.6.3 Conclusions

The effect of the location of the dimple on the shock performance of a small form factor drive was investigated. Simulations were carried out using the CML Dynamic L/UL/S simulator using a shock pulse. The results of the study are summarized in Fig. 6.35. We see that maximum shock resistance is realized when the dimple is located at $(-0.13, 0.015)$. It was shown that moving the dimple towards the leading edge of the slider improves shock performance when the mechanism of shock failure is load beam-flexure impact. However if the dimple is moved too far, the mechanism of shock failure changes and shock performance degrades.

6.7   Discussion

In this chapter we presented a methodology for simulating the shock response in hard disk drives. This is followed by results from parametric studies on investigating the effect of the nature of shock (pulse-width) and suspension design (specifically the location of the dimple on the slider).

6.8 Tables

Table 6.1: Disk parameters for shock studies

Parameter	Value
Inner diameter	3.45 mm
Outer diameter	13.68 mm
Thickness	0.38 mm
Young's modulus	75.0 GPa
Density	2.71 g/mm^3
Poisson's ratio	0.3

Table 6.2: Slider parameters for shock studies

Parameter	Value
Drive form factor	1"
Gram load	1.25 g
RPM	3600
Steady state fly height (OD)	6.24 nm
Steady state pitch	58.3 μrad
Steady state roll	-2.6 μrad
Operating PSA	2.5 mrad
Operating RSA	0.0 mrad

6.9   Figures

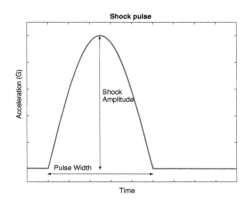

Figure 6.1: Half-sine shock pulse

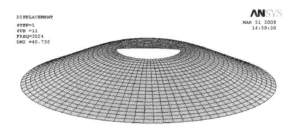

Figure 6.2: The disk umbrella mode

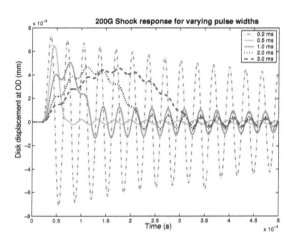

Figure 6.3: Disk response for 200G shock of varying pulse widths

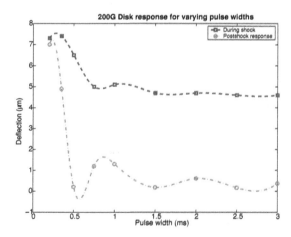

Figure 6.4: Maximum disk deflection for 200G shock of varying pulse widths

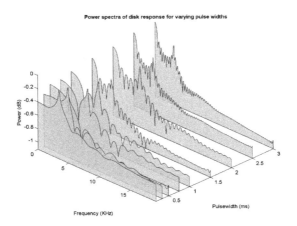

Figure 6.5: Disk response spectra for 200G shock of varying pulse widths

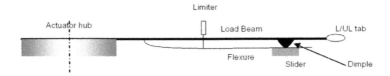

Figure 6.6: Suspension schematic

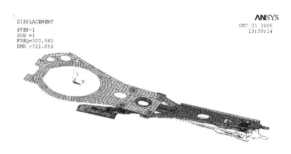

Figure 6.7: Suspension first bending mode

Figure 6.8: Suspension first torsion mode

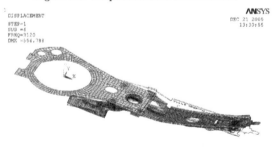

Figure 6.9: Suspension second bending mode

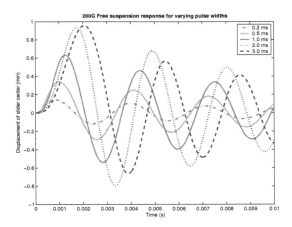

Figure 6.10: Suspension response to a 200G shock of varying pulse widths

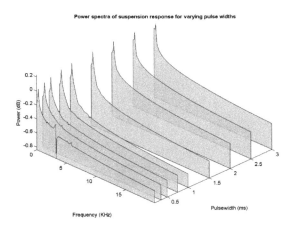

Figure 6.11: Suspension response spectra for a 200G shock of varying pulse widths

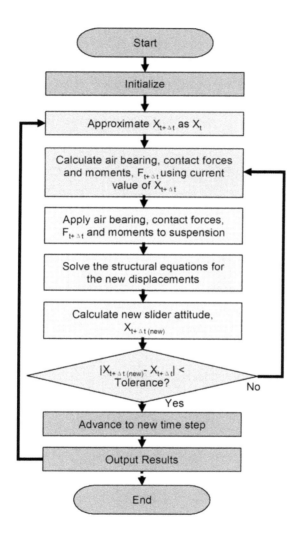

Figure 6.12: Structural Air-bearing coupling scheme

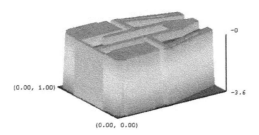

Figure 6.13: Slider Design

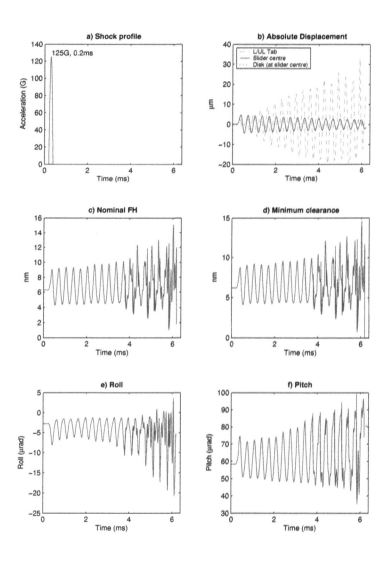

Figure 6.14: Slider Response for 125G, 0.2ms shock

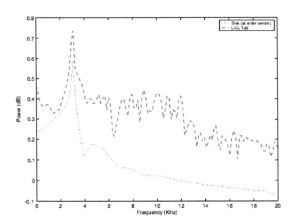

Figure 6.15: Frequency spectra of load beam and disk response to 125G, 0.2ms shock

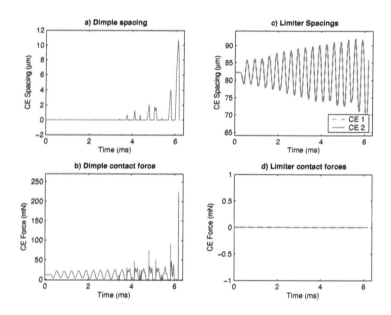

Figure 6.16: Dimple, limiter status for 125G, 0.2ms shock

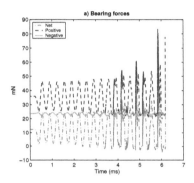

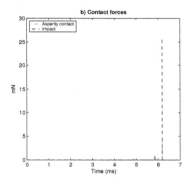

Figure 6.17: Air bearing and contact forces for 125G, 0.2ms shock

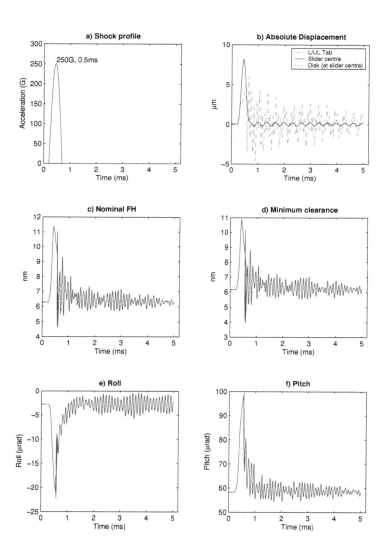

Figure 6.18: Slider Response for 250G, 0.5ms shock

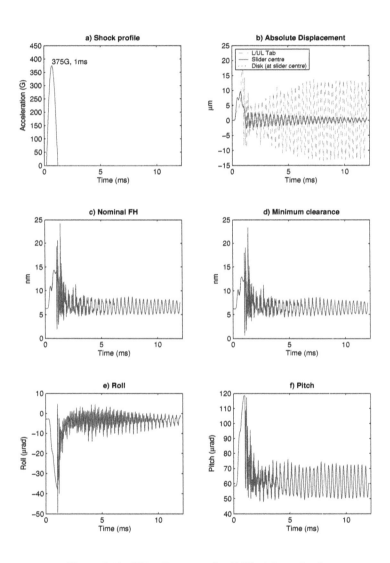

Figure 6.19: Slider Response for 375G, 1.0 ms shock

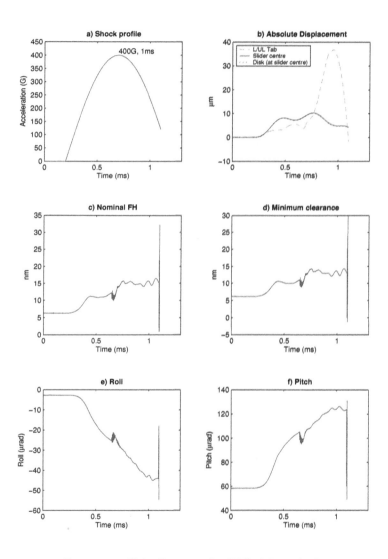

Figure 6.20: Slider Response for 400G, 1.0 ms shock

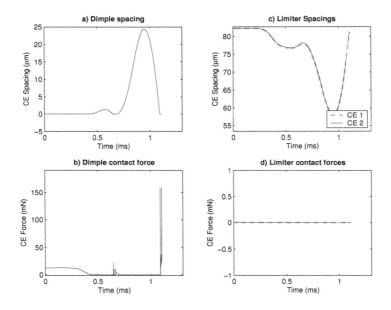

Figure 6.21: Dimple, limiter status for 400G, 1.0 ms shock

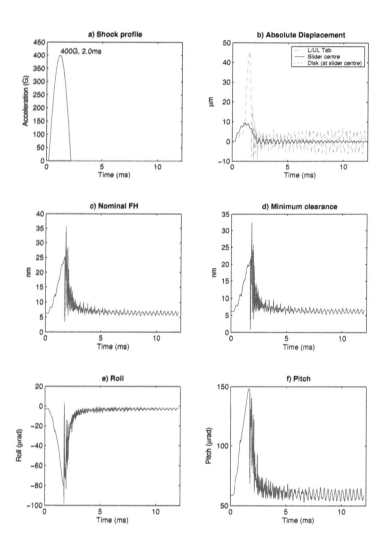

Figure 6.22: Slider Response for 400G, 2.0 ms shock

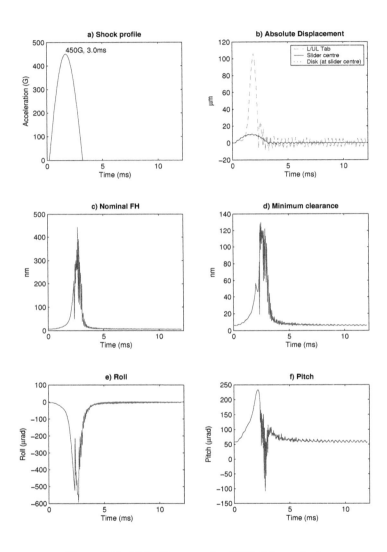

Figure 6.23: Slider Response for 450G, 3.0 ms shock

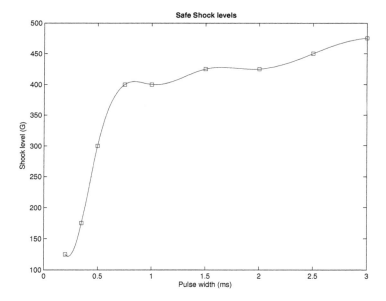

Figure 6.24: Safe shock levels for varying pulse widths

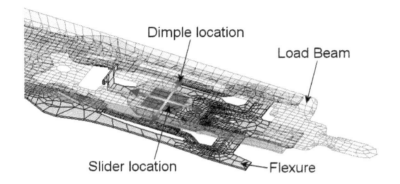

Figure 6.25: Suspension system

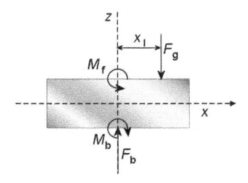

Figure 6.26: Slider free-body diagram

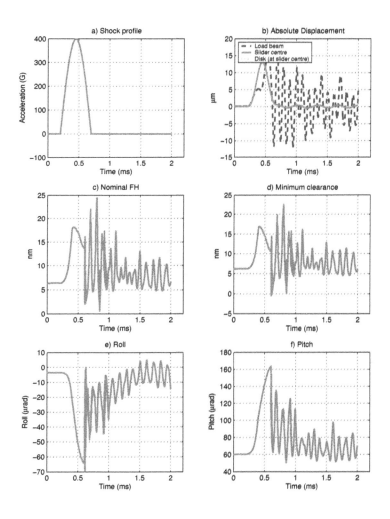

Figure 6.27: Slider attitude for BC 400G shock

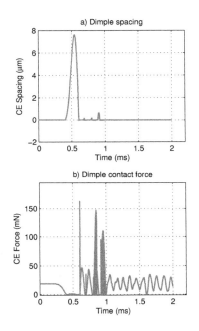

Figure 6.28: Dimple separation and contact force for BC 400G shock

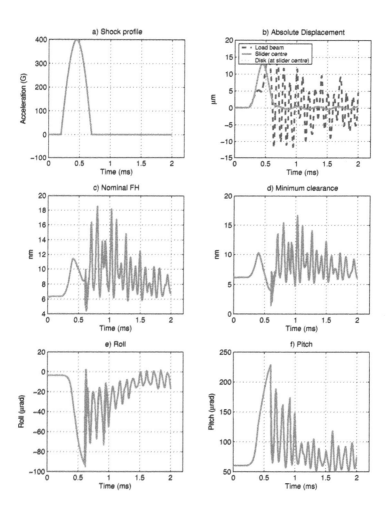

Figure 6.29: Slider attitude for U1 400G shock

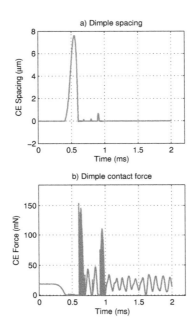

Figure 6.30: Dimple separation and contact force for U1 400G shock

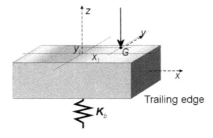

Figure 6.31: Linearized system

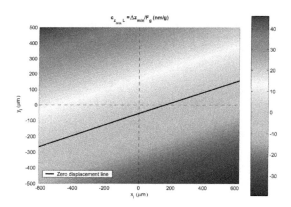

Figure 6.32: $c_{z_{min}G}$ variation along x_l and y_l

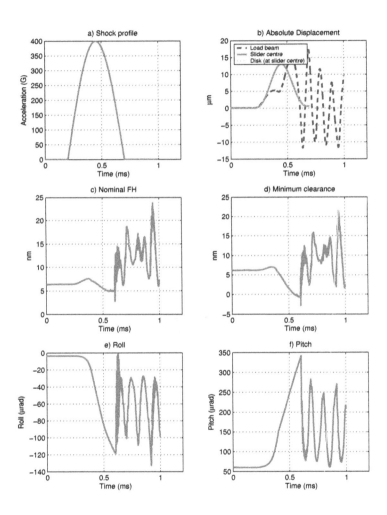

Figure 6.33: Slider attitude for U2 400G shock

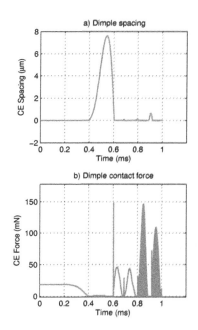

Figure 6.34: Dimple separation and contact force for U2 400G shock

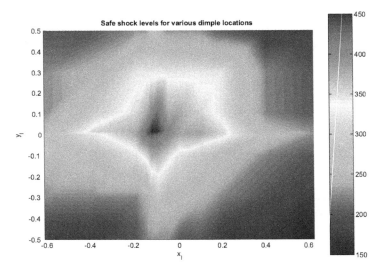

Figure 6.35: Safe shock levels for various dimple locations

Chapter 7

Simulating the Load/Unload Process

7.1 Introduction

Traditionally HDDs have prevented power interruptions from shutting the drive down with its heads landing in the data zone by moving the heads to a special 'landing zone'. A landing zone is an area of the platter usually near its inner diameter (ID), where no data is stored. This area is also called the Contact Start/Stop (CSS) zone. Although in CSS drives the sliders are designed to survive many landings and takeoffs from the disk, the wear on the slider and the disk eventually takes its toll and causes the drive to fail. Most manufacturers design the sliders to survive 50,000 contact cycles before the probability of failure on startup exceeds 50%. Around 1995 IBM pioneered laser texturing of landing zones producing an array of smooth nanometer-scale 'bumps' in the landing zone. This technology is still in use today, predominantly in desktop and enterprise (3.5 inch) drives. However, CSS technology is still prone to increased stiction (the tendency of the heads to stick to the disk surface due to intermolecular and meniscus forces) as well as wear due to the rubbing of head-disk at the landing zones.

Load/Unload technology relies on the heads being lifted off the disk onto a ramp, thus eliminating the risks of wear and stiction altogether. Load/Unload (L/UL) technology today is finding widespread application not only in small form factor drives, but also in desktop and high-end server drives. Although the first application of load/unload technology was as early as the late 1950s [IBM corporation, 2008] the ramp load/unload technology of today first appeared in mobile drives in the early 1990s. The ramp load/unload system is now well established as the standard for almost all segments of small form factor drives: from 2.5" mobile drives to the 1"-0.8" drives, which today find widespread application in a host of consumer electronic devices. Load/unload technology offers many advantages over the traditional CSS technology. One of the most important advantages is the much greater shock resistance in the non-operational state. Today's drives equipped with L/UL systems can withstand shocks, which would fatally damage the air-bearing slider and the disk in CSS systems. And with the emergence of smaller form factor drives such as, the 1" and the 0.8" drives which find widespread application in shock-prone environments like MP3 players and cameras, shock resistance is perhaps the biggest advantage of L/UL technology. Another advantage of L/UL is that of avoiding the problem of stiction, which CSS systems are inherently prone to. This leads to lower power consumption as well as lower wear and debris for the ABS. L/UL technology also has the potential to do away with CSS landing zones, and with efforts being made to load/unload on data tracks, this valuable space on

the disk can be recovered to increase the storage capacity of the drive. The main design

objectives of L/UL are to avoid or at least minimize the occurrence of slider-disk contacts

leading to media and head damage, small ramp forces such that ramp wear is minimized,

and a smooth and short unloading process.

7.2   Previous work

In the period from 1988 to 2003, many pioneering studies [Zeng and Bogy, 2000a,

1999a,b, 2000b; Zeng, Chapin, and Bogy, 1999; Jeong and Bogy, 1990, 1991, 1993; Chapin

and Bogy, 2000; Yamada and Bogy, 1988] on L/UL were carried out at the Computer

Mechanics Laboratory (CML) at U.C. Berkeley. It is found that the suspension model is

critical in the simulations. Zeng and Bogy [2000a] implemented a reduced suspension model

for the purpose of load unload simulations. They included four degrees of freedom for the

suspension: one for the lift (L/UL) tab and three for the slider z-height, pitch and the roll.

They also incorporated suspension contact nonlinearities at the dimple and the limiters by

including different suspension *states*, where in a different suspension stiffness matrix would

be used for each of the suspension states (i.e. dimple closed, limiters open; dimple open,

limiters open; dimple open, limiters closed; etc.). This model was used in various studies

to carry out parametric and slider design studies [Zeng and Bogy, 1999a,b, 2000b]. The

performance of the model was also compared with experimental measurements [Zeng et al.,

1999; Chapin and Bogy, 2000]. This model was perhaps the most accurate L/UL model

previously developed. However it suffered from some limitations in terms of accurately capturing suspension dynamics.

In this chapter we present a new L/UL simulator with more sophisticated and complete modeling for the suspension (as described in Chapter 2). The new simulator allows us to model the actuator swinging motion during the load and unload processes so that these processes can be more realistically simulated. This new simulator also allows us to simulate the loading and unloading processes much more accurately without any significant loss in computational efficiency as compared to the simulator developed by Zeng and Bogy [2000a]. In this chapter we present the basic methodology employed to simulate the loading and unloading processes as well as some results from this new simulator and compare these results with those predicted by the previous L/UL Simulator developed at CML [Zeng and Bogy, 2000a].

7.3   Methodology

The CML L/UL/S simulator discussed in chapter 6 is used to carry out the L/UL simulations carried out in this chapter.

7.4   Numerical Simulations

Simulations were carried out for two loading and unloading processes, and the results were compared with those of the previous version of the CML L/UL simulator. The simulations were carried out using a slider from a currently popular 1" drive shown in Fig. 7.2. The operation parameters for this slider suspension system are listed in Table 7.1.

The new simulator models actuator rotation over a user defined ramp profile. However since the 4-DOF model simulator models the loading and unloading processes as a constant velocity load and unload, a flat ramp profile of inclination 27.7 degrees was used here in the new simulator for comparison of the two simulators. Also the loading and unloading processes are coupled with a track seek motion in the 4-DOF simulator to model the effect of the actuator motion.

Table 7.1: Slider parameters for load/unload comparison studies

Parameter	Value
Drive form factor	1"
Gram load	1.25 g
RPM	3600
Steady state fly-height (OD)	5.5 nm
Steady state pitch	95 μrad
Steady state roll	-5 μrad
Operating PSA	10.0 mrad
Operating RSA	0.0 mrad

7.4.1 Unloading process

The unloading process is simulated using the new as well as the previous versions of the CML L/UL simulators. The first unloading process simulated is for a vertical unloading velocity of 44.44 mm/s (this corresponds to turning the actuator at a constant angular velocity of 4 rad/s). The simulation results are plotted in Figs. 7.3-7.6. Figure 7.3 has several plots showing various components of the attitude of the slider for the two simulators; the slider center displacements in a), the nominal fly heights in b), the minimum clearances in c) and the rolls and the pitches in subplots d) and e) respectively. We see that the unloading results predicted by the two simulators are very similar, with both the simulators predicting

that unloading occurs at about 3.1 ms. However, we notice that more suspension oscillations

as well as lesser damping are predicted by the FE suspension model based simulator (using

similar damping parameters), especially in the z and the pitch directions. Fig. 7.4 plots

the forces during the unloading process. The force curves show remarkable agreement, with

both the simulators predicting slider disk impact at about 2.5 ms and 3 ms. In Fig. 7.5

we plot the ramp and the tab positions in a) and ramp contact forces for both of the

simulators in b). We see that the ramp contact forces follow the same trend in the two

simulations, however, in the FE based simulator we see a much richer frequency content as

well as the effect of oscillations during the unloading process. In Fig. 7.6, we plot the status

of the contact elements in the FE based simulator, which gives us information regarding

the dimple contact forces and the limiter forces. In a), we see how the dimple separates

and following the unloading process, how the flexure oscillates and impacts the load beam.

These contact forces are plotted in b), where we see that the maximum magnitude of these

forces exceeds 60 mN, which would likely lead to wear of the dimple and also generate debris

inside the drive. In c) and d) we see that the limiters do not engage during this particular

unloading process. The two curves correspond to the two contact elements on both sides of

the hammer head limiter. This data is not compared to the 4-DOF based simulator, since

that simulator is unable to generate these results. The second unloading simulation was

for an unloading velocity of 88.89 mm/s, achieved by doubling the angular velocity of the

actuator to 8 rad/s. The slider attitudes for this process are plotted in Fig. 7.7. Again the

results from the two simulators are in good agreement. Fig. 7.8 shows plots of the forces for the second unloading process. The forces are in good agreement, however there is a slight difference after about 1.8 ms, which is the time when the limiters engage. This difference can be attributed to the fact that the new simulator describes the actual impact between the limiters and the load beam using contact elements rather than modeling the limiter engagement solely as a change in stiffness of the suspension. In Fig. 7.9 a) we again plot the ramp and tab positions as a function of time. In b) we plot a comparison of the ramp forces. The results are similar to those discussed for the previous unloading process. In Fig. 7.10, we plot the contact element behavior. Subplots a) and b) show the dimple spacing and contact forces. Here we see that the contact forces at the dimple exceed 80 mN, which may be harmful for the drive in terms of particle contamination. In c) we plot the limiter spacing, where we see that the limiter closes at about 2 ms, after which the slider quickly unloads from the disk.

7.4.2 Loading process

The loading process was first simulated using a vertical loading velocity of 50 mm/s. The results for the comparison of this loading process between the two simulators are plotted in Figs. 7.11-7.14. Figure 7.11 shows plots of the slider attitude as predicted by the two simulators. The various quantities plotted are the slider center displacements in a), the nominal fly heights in b), minimum clearances in c) and rolls and pitches in d) and e) respectively. We see that the HGA moves down the ramp, and the slider reaches the disk at about 3 ms. We see in Fig. 7.12 that the processes predicted by the two simulators are qualitatively and quantitatively very similar in the slider behavior as well as the development of the air bearing. There are no impacts during the loading process and the magnitudes of the asperity contact forces are also almost the same. In Fig. 7.13, we plot the ramp and tab positions in a) as well as the ramp contact forces in b). In Fig. 7.14, we plot the contact element data. In a) we see that the dimple is initially in the open position. This is because even though there is a dimple preload, the bending of the load beam due to the ramp negates the effect of the preload in the parked state. During the course of the loading process, oscillations cause the flexure to strike the load beam until it finally closes at about 2.4 ms as the slider begins to load onto the disk through the air bearing. For the second loading simulation case, we increase the loading velocity to 66.67 mm/s, using the same ramp profile, but increasing the actuator angular velocity to 8 rad/s. Again we observe that the attitudes predicted by the two simulators, as plotted in Fig. 7.15, are quite similar. In this case, however we see that the 4-DOF model predicts slider disk impact at about 2.5 ms. In c) in the inset zoomed plot, we can see that there is an impact at about 2.5 ms. Also since the pitch is negative, it is the front of the slider that hits the disk. This can also be seen from the force plots in Fig. 7.16. In c) we see an impact between the slider and the disk at about 2.5 ms when the air bearing has not yet fully developed. This seems to be the result of strong oscillations of the slider in the pitch direction. In the 4-DOF model based simulator, all of the suspension mass is transferred to the slider as the 'effective mass' of the slider. Hence the slider has the same effective mass irrespective of whether the dimple is closed or open, whereas in reality this would depend on whether the dimple is closed or open. Thus we see that when the dimple closes during the loading process at 2 ms as seen in Fig. 7.18, the magnitude of the pitch oscillations reduces as some of the energy is transferred to the load beam for the FE model based simulator, whereas for the 4-DOF simulator, the oscillation magnitude remains the same. This leads to the slider-disk impact. Increasing the loading speed, we observe similar behavior with the 4-DOF model based simulator, predicting contacts which may not have occurred.

7.4.3 Frequency response

Figure 7.19 shows a comparison of the frequency contents between the two simulators for a typical loading process. In a), b) and c), we plot the spectra for the 4-DOF based simulator and in d), e) and f), we plot the spectra for the FE based simulator. In a) we observe two distinct peaks labeled 1 and 2. Peak 1 corresponds to the first bending mode of the suspension at 320 Hz, which was used to calculate the effective mass of the slider in the 4-DOF suspension model. The second peak, at about 800 Hz, corresponds to the bending frequency of the system when the dimple is open. Since the effective mass of the slider is unchanged in this state, this is actually a 'false' frequency, a figment of the mathematical modeling. In b), which shows plots of the pitch curve, we again see these two frequency peaks. In c) where the roll frequency is plotted we again observe the false frequency at 800 Hz, which is seen in roll as a result of the coupling between roll and bending (since the stiffness matrix is not diagonal). We also observe a sharp peak 3 at 2700 Hz which corresponds to the first torsion mode of the suspension, which was used to calculate the effective mass in the roll direction in the 4-DOF model. For the FE-model based simulator, we observe much richer frequency spectra, as well as no false frequencies. In d), we observe various peaks labeled 1, 2, 3, 4 and 5 which correspond to the first bending (320 Hz), two flexure bending modes (dimple open, 1160 Hz), flexure bending mode (dimple closed, 1700 Hz), a third flexure bending mode (dimple open, 2100 Hz) and two flexure-load beam coupled bending modes (dimple open, 3220 Hz and 3180 Hz dimple closed). In e) which plots the pitch frequency, we see peaks 1,2,4,5 and in addition peaks 6 and 7 which correspond to a bending-torsion coupled mode (4056 Hz) and a load beam-second bending torsion coupled mode of the suspension (5900 Hz). For the roll, plotted in f), we observe peaks at 8, 6 and 9. Here 8 is the first torsion mode (2700 Hz) of the suspension.

7.5   Conclusion

We discuss the methodology developed for carrying out load/unload simulations. We compared simulation results obtained using the new simulator with those based on the CML 4-DOF model for two loading and unloading cases. We found that the results are in excellent agreement for unloading and for the slower loading processes. However for faster loading processes, we found slider-disk contacts in results obtained using the 4-DOF simulator which were not predicted by the FE based simulator. We explained these false contacts as resulting from the same effective mass being used for the suspension for all suspension states. Finally we also compared the spectra of various parameters and found the existence of false frequencies in the 4-DOF based simulator as well as a much richer modal representation for the FE model based simulator. This makes the FE based L/UL simulator much more useful in the simulation of faster loading and unloading processes in the presence of strong disturbances as well as other phenomena such as shock and vibration.

7.6 Figures

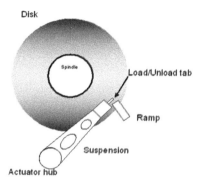

Figure 7.1: Schematic of a 1" drive

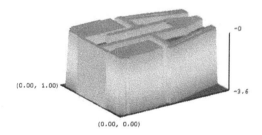

Figure 7.2: Slider used for simulations

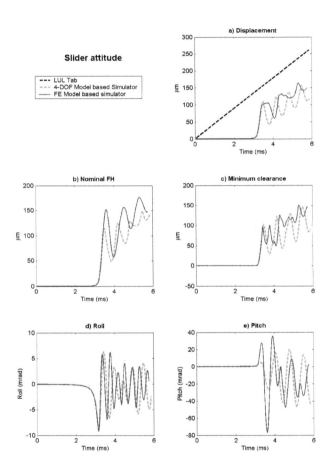

Figure 7.3: Slider attitude history comparison during the unloading process (44.4 mm/s)

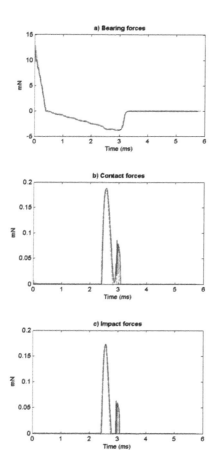

Figure 7.4: Force history comparison during the unloading process (44.4 mm/s)

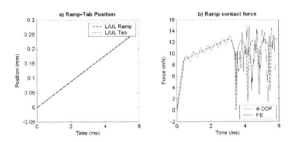

Figure 7.5: Ramp-tab position and ramp contact force during the loading process (44.4 mm/s)

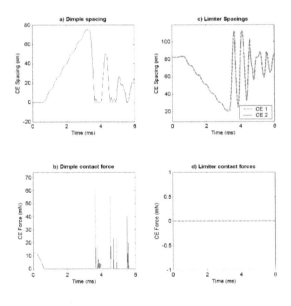

Figure 7.6: Dimple and limiter contact status during the loading process (44.4 mm/s)

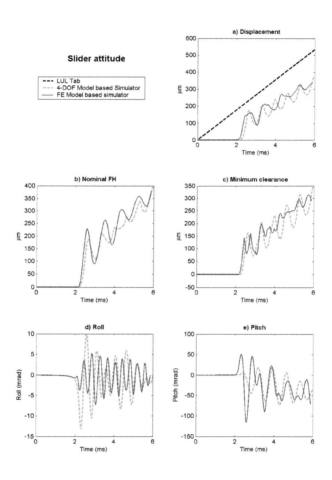

Figure 7.7: Slider attitude history comparison during the loading process (88.9 mm/s)

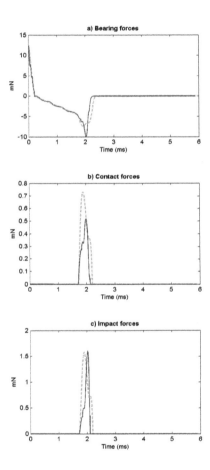

Figure 7.8: Force history comparison during the unloading process (88.9 mm/s)

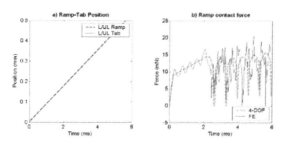

Figure 7.9: Ramp-tab position and ramp contact force during the loading process (88.9 mm/s)

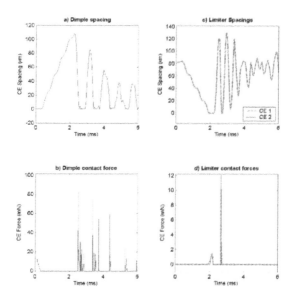

Figure 7.10: Dimple and limiter contact status during the loading process (88.9 mm/s)

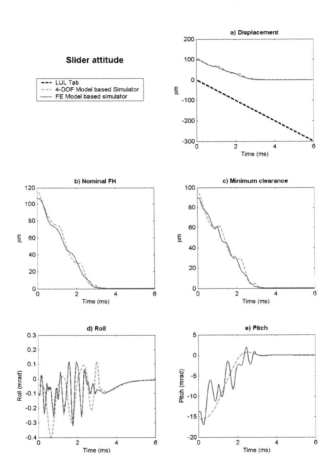

Figure 7.11: Slider attitude history comparison during the loading process (50 mm/s)

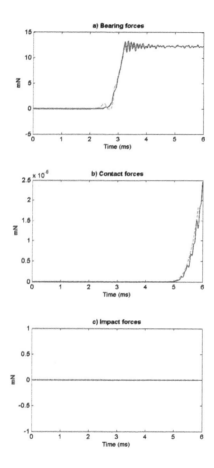

Figure 7.12: Force history comparison during the loading process (50 mm/s)

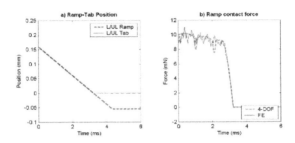

Figure 7.13: Ramp-tab position and ramp contact force during the loading process (50 mm/s)

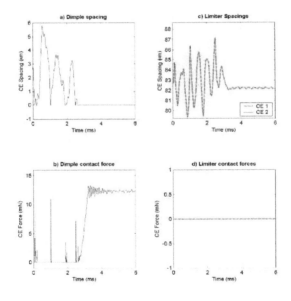

Figure 7.14: Dimple and limiter contact status during the loading process (50 mm/s)

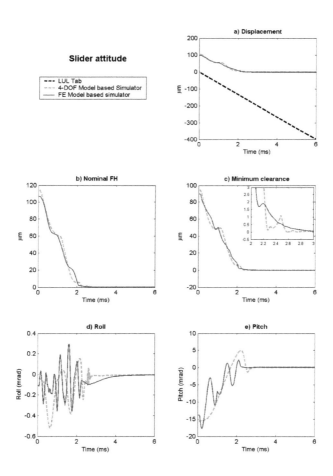

Figure 7.15: Slider attitude history comparison during the loading process (66.7 mm/s)

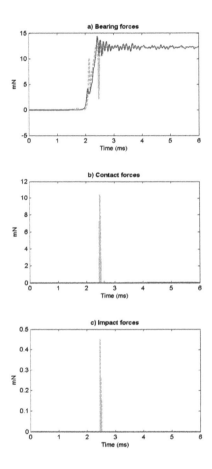

Figure 7.16: Force history comparison during the loading process (66.7 mm/s)

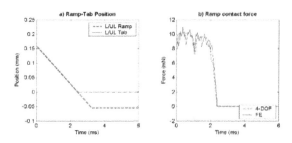

Figure 7.17: Ramp-tab position and ramp contact force during the loading process (66.7 mm/s)

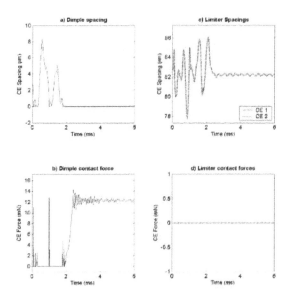

Figure 7.18: Dimple and limiter contact status during the loading process (66.7 mm/s)

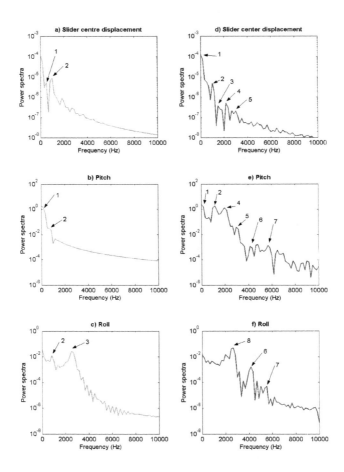

Figure 7.19: Frequency spectra of a typical L/UL process for 4-DOF (a, b, c) and FE based (d, e, f) simulators

Chapter 8

Conclusion and Future Work

8.1 Conclusion

With the increased usage of hard disk drives in consumer electronic devices such as digital video recorders and personal audio players, as well as in laptops and ultra-portable computers, there is a great need for hard disk drives, now more than ever, to perform under the most adverse conditions with high reliability. Although the improvements made by the industry over the past few decades in this area have been phenomenal, there is a continuous necessity for the improvement of reliability of the drive and to make the head-disk interface more robust. Also with the continual increase in the areal density of drives, the head-media mechanical spacing has to be made smaller or the slider flown closer to the disk. This has necessitated the need for complex slider designs with steep wall profiles and intricate etch designs to meet the demands of low fly-height and high reliability.

The research presented in this dissertation is hopefully a contribution in meeting the demands of tomorrow. Starting from the basic equations of motion for the suspension/air-bearing/disk system, we develop a methodology to solve the coupled equations of motion for the fluid-structural system.

To model the structural components of the system, we use finite element models. To improve computational efficiency, dynamic reduction is carried out to reduce the degrees of freedom of the system. The nonlinear contact constraints are imposed on the reduced models using a Lagrange multiplier formulation and impact/release conditions.

In order to model the impact between the head and the disk, we propose a new boundary element formulation for the quasi-static head-disk indentation problem. The model is compared with exact solutions available in the literature and found to be in good agreement. We also a propose a method to speed up the calculations by limiting the *radius of effect*. The method is implemented for modeling head-disk impacts in our dynamic simulator.

The air-bearing is perhaps the most critical component of our system. The air-bearing is modeled using the Reynolds equation which is generalized to account for rarefaction effects of air, due to the extremely small spacing between in the air bearing. We employ a finite element method with SUPG stabilization to solve the generalized Reynolds equation. We derive the weak form and linearize the resulting equations first with respect to pressure to solve the *forward problem*, which leads to a series of banded linear systems to find the solution for the pressure field. To solve the *inverse problem*, we first present the traditional method, whereby the flying attitude of the slider is obtained by solving a series of forward problems. Expressions for the stiffness and damping of the air bearing are derived. This method requires two levels of iterations, the first to solve the equations of motion for the slider attitude, and the second to solve the Reynolds equation for the pressure field. An alternative linearization is suggested whereby pressure and slider attitude are both treated as variables. This leads to a series of arrowhead linear systems to solve for the pressure and slider attitude together. In this case only one level of iterations is present. We next investigate refinement strategies to obtain optimal meshes for solving the steady state problem. Pressure gradients and pressure flux jumps are found to work very well for all cases considered. The air-bearing formulation is next extended to the general time-dependent case. Expressions for the algorithmic tangent stiffness are derived. The method is implemented and tested.

Finally we employ the framework developed to solve practical problems of shock and load/unload. Parametric studies on shock and load/unload are presented which have been carried out using a previous implementation of the above framework (using a finite volume formulation [Hu and Bogy, 1995] for the air bearing instead of the finite element formulation discussed here). For shock, we investigate the mechanism of failure for various *types* of shock which are characterized by varying the pulse widths of the shock pulse for a 1" drive. We plot the safe shock levels for varying pulse widths. We find that, as expected, shorter pulse widths are more critical than longer pulse widths in terms of shock magnitudes. We also investigate the effect of varying the location of the dimple on the suspension on the shock performance of a the same drive. It is found that instead of locating the dimple at the center of the slider, if the dimple is offset a few hundred microns towards the leading edge, shock resistance can be improved significantly. Finally we carry out load/unload simulations and compare our methodology with a simpler one previously developed at CML. The differences in results predicted by the two methods are explained on the basis of models used by them.

8.2 Future Work

The simulation tools developed as part of this research have been made available for use to CML industry members. However improvements and additions to the methodology and program will continue to be developed at CML.

The modeling of the structural components in our model has been strictly component level, i.e. only individual components have been modeled, without the entire structure of the drive such as the bas plate and spindle motor/bearing and bearings for the actuator. It has been observed in experiments that the stiffness of these components can play a role in the dynamics of the system, especially during shock. Some researchers in industry have used the CML L/UL/S simulator developed herein to numerically investigate the effect of the base plate and spindle bearing and found that they can affect the shock performance of the hard disk drive significantly (private discussions). However instead of such ad-hoc approaches a more complete framework can be developed for including the complete disk drive in terms of structural modeling.

In our method, we have modeled the disk as stationary. The assumption of the response of the spinning disk being closely approximated by a stationary disk is only valid for axisymmetric shocks. For non-axisymmetric shocks, the rotation of the disk becomes important and needs to be considered.

Another direction of future research could be in terms of validation of the simulation code developed herein with experiments. Although word-of-mouth figures on shock limits from experimental investigations from industry lie in the same ball park as those predicted from simulations, a formal validation study correlating experiments and simulations would lend more credibility to the simulation results.

188

Bibliography

ANSYS. *ANSYS 8.1 Documentation*, 2005.

P. Bhargava and D. B. Bogy. The CML Finite Element Dynamic Simulator. Technical Report 2008-003, CML, University of California, Berkeley, 2007a.

P. Bhargava and D. B. Bogy. Numerical Simulation of Load/Unload in Small Form Factor Hard Disk Drives. Technical Report 2005-011, CML, University of California, Berkeley, 2005.

P. Bhargava and D. B. Bogy. Numerical Simulation of Operational-Shock in Small Form Factor Hard Disk Drives. *Journal of Tribology*, 129(1):153–160, 2007b.

P. Bhargava and D. B. Bogy. Effect of Shock Pulse Width on the Shock Response of Small Form Factor Disk Drives. *Microsystem Technologies*, 13(8-10):1107–1115, 2007c.

P. Bhargava and D. B. Bogy. The CML Dynamic Load/Unload/Shock Simulator (version 5.1). Technical Report 2006-009, CML, University of California, Berkeley, 2006.

A. N. Brooks and T. J. R. Hughes. Streamline Upwind/Petrov-Galerkin Formulations for Convection Dominated Flows with particular emphasis on the Incompressible Navier-

Stokes Equations. *Computer methods in applied mechanics and engineering*, 32:199–259, 1982.

V. Castelli and J. Pirvics. Review of Numerical Methods in Gas Bearing Film Analysis. *ASME Journal of Lubrication Technology*, 90:777–792, 1968.

E. Cha and D.B. Bogy. A Numerical Scheme for Static and Dynamic Simulation of Sub-ambient Pressure Shaped Rail Sliders. *Journal of Tribology, Transactions of the ASME*, 117:36–43, 1995.

M. Chapin and D. B. Bogy. Air Bearing Force Measurements of Pico Negative Pressure Sliders during Dynamic Unload. *ASME Journal of Tribology*, 122(4):771–775, 2000.

L. Chen, Y. Hu, and D. B. Bogy. The CML Air Bearing Dynamic Simulator (Version 4.21). Technical Report 1998-004, CML, University of California, Berkeley, 1998.

D.H. Choi and S.J. Yoon. Static Analysis of Flying Characteristics of the Head Slider by using an Optimization Technique. *ASME Journal of Tribology*, 116:90–94, 1994.

E. D. Daniel, C. D. Mee, and M. H. Clark. *Magnetic Recording - The First 100 Years*. IEEE Press, New York, 1999.

L. C. Dutto. The Effect of Ordering on Preconditioned GMRES Algorithm, for solving the Compressible Navier-Stokes Equations. *International Journal for Numerical Methods in Engineering*, 36:457–497, 1993.

J. R. Edwards. Finite Element Analysis of the Shock Response and Head Slap Behavior of a Hard Disk Drive. *IEEE Trans. of Magnetics*, 35:863–867, 1999.

G. Fu and A. Chandra. Normal Indentation of Elastic Half-Space with a Rigid Frictionless Axisymmetric Punch. *Transactions of the ASME*, 69:142–146, 2002.

S. Fukui and R. Kaneko. A Database for Interpolation of Poiseuille Flow Rates for High Knudsen Number Lubrication Problems. *Trans. ASME Jounal of Tribology*, 112:78–83, 1990.

C. Garcia-Suarez, D.B. Bogy, and F.E. Talke. Use of an Upwind Finite Element Scheme for Air Bearing Calculations. *Tribology and Mechanics of Magnetic Storage Systems*, 1: 90–96, 1984.

J.A. Greenwood and J.B.P. Williamson. Contact of Nominally Flat Surfaces. *Proceedings of the Royal Society of London*, 295(1442):300–319., 1966.

R. Grisso and D. B. Bogy. The CML Air Bearing Design Program (CMLAir32) Version 5 User Manual. Technical Report 1999-024, CML, University of California, Berkeley, 1999.

V. Gupta. *Air Bearing Slider Dynamics and Stability in Hard Disk Drives*. PhD thesis, University of California, Berkeley, 2007.

V. Gupta and D. B. Bogy. Effect of Intermolecular Forces on the Static and Dynamic Performance of Air Bearing Sliders: Parts I, II. Technical Report 2004-001,002, CML, University of California, Berkeley, 2004.

R.J. Guyan. Reduction of Stiffness and Mass Matrices. *AIAA Journal*, 3:310, 1965.

E. Hairer and G. Wanner. *Solving Ordinary Differential Equations II*. Springer, 1996.

J. C. Harrison and M. D. Mundt. Flying Height Response to Mechanical Shock during

   Operation of a Magnetic Hard Drive. *ASME Journal of Tribology*, 122:260–263, 2000.

F. Hendricks. A Design Tool for Steady State Gas Bearings using Finite Elements, the

   APL Language and Delaunay Triangulation. *ASLE Tribology and Mechanics of Magnetic*

   Storage Systems, 5:124–129, 1988.

Hitachi GST. http://www.hitachigst.com/portal/site/en/, March 2008.

P. Holani and S. Müftü. An Adaptive Finite Element Strategy for Analysis of Air Lubrica-

   tion in the Head-Disk Interface of a Hard Disk Drive. *Revue Europenne des lment Finis*,

   14:155–180, 2005.

Y. Hu and D. B. Bogy. The CML Air Bearing Dynamic Simulator. Technical Report

   1995-011, CML, University of California, Berkeley, 1995.

T.J.R. Hughes, R.L. Taylor, J.L. Sackman, A. Curnier, and W. Kanoknukulchai. A Finite

   Element Method for a Class of Contact-Impact Problems. *Computer methods in applied*

   mechanics and engineering, 8:249–276, 1976.

IBM corporation. http://storage.ibm.com, March 2008.

A. Iserles. *A First Course in the Numerical Analysis of Differential Equations*. Cambridge

   University Press, 1996.

E. M. Jayson, J. Murphy, P. W. Smith, and F. E. Talke. Shock Modeling of the Head-Media

   Interface in an Operational Hard Disk Drive. *IEEE Trans. of Magnetics*, 39:2429–2432,

   2003.

T. G. Jeong and D. B. Bogy. Numerical Simulation of Dynamic Loading in Hard Disk Drives. *ASME Journal of Tribology*, 115:370–375, 1993.

T.G. Jeong and D. B. Bogy. Slider Disk Interactions during the Load-Unload Process. *IEEE Transactions on Magnetics*, 26:2490–2492, 1990.

T.G. Jeong and D. B. Bogy. Measurements of Slider-Disk Contacts during Dynamics Load-Unload. *IEEE Transactions on Magnetics*, 27:5073–5075, 1991.

Z. W. Jiang, K. Takashima, and S. Chonan. Shock Proof Design of Head Disk Assembly Subjected to Impulsive Excitation. *JSME International Journal*, 38:411–419, 1995.

T. Kouhei, T. Yamada, Y. Keroba, and K. Aruga. A Study of Head-Disk Interface Shock Resistance. *IEEE Trans. of Magnetics*, 31:3006–3008, 1995.

M. Kubo, Y. Ohtsubo, N. Kawashima, and H. Marumo. Finite Element Solution for the Rarefied Gas Lubrication Problem. *Journal of Tribology, Transactions of the ASME*, 110 (2):335–341, 1988.

S. Kumar, V. Khanna, and M. Sri-Jayantha. A Study of the Head Disk Interface Shock Failure. In *The 6th MMM-Intermag Conference*, 1994.

S. Lu. *Numerical Simulation of Slider Air Bearings*. PhD thesis, University of California, Berkeley, 1997.

A. I. Lur'e. *Theory of Elasticity*. Foundations of Engineering Mechanics. Springer, 2005.

D.K. Miu and D.B. Bogy. Dynamics of Gas-Lubricated Slider Bearing in Magnetic Recording Disk Files Part II: Numerical Simulation. *Journal of Tribology, Transactions of the ASME*, 108:589–593, 1986.

K. Ono. Dynamic Characteristics of the Air-Lubricated Slider Bearing for Non-Contact Magnetic Recording. *ASME Journal of Lubrication Technology*, 97:250–260, 1972.

S. Patankar. *Numerical Heat Transfer and Fluid Flow*. Hemisphere Series on Computational Methods in Mechanics and Thermal Science. Taylor and Francis, 1980.

M. Paz and W. E. Leigh. *Structural Dynamics: Theory and Computation*. Structural dynamics. Springer, 2004.

J.P. Peng and C.E. Hardie. A Finite Element Scheme for determining the Shaped Rail Slider Flying Characteristics with Experimental Confirmation. *Trans. ASME Jounal of Tribology*, 117:358–364, 1995.

V. Ponnaganti. *Dynamics of Head-Disk Interaction in Magnetic Recording*. PhD thesis, Stanford University, 1986.

A. Quarteroni, R. Sacco, and F. Saleri. *Numerical Analysis*. Springer, 2000.

O.J. Ruiz and D.B. Bogy. A Numerical Simulation of the Head-Disk Assembly in Magnetic Hard Disk Files: Part 1 - Component Models. *Journal of Tribology, Transactions of the ASME*, 112:593–602, 1990.

Y. Saad. *Iterative Methods for Sparse Linear Systems*. SIAM, 2003.

Seagate Technologies. http://www.seagate.com, March 2008.

J. R. Shewchuk. Triangle: Engineering a 2D Quality Mesh Generator and Delaunay Triangulator. In Ming C. Lin and Dinesh Manocha, editors, *Applied Computational Geometry: Towards Geometric Engineering*, volume 1148 of *Lecture Notes in Computer Science*, pages 203–222. Springer-Verlag, May 1996. From the First ACM Workshop on Applied Computational Geometry.

P. W. Smith, M. H. Wahl, and F. E. Talke. Accelerated Natural Convergence for Pivoted Slider Bearings. *Tribology Transactions*, 38(3):595–600, 1995.

T. Tang. Dynamics of Air-Lubricated Slider Bearings for Non-Contact Magnetic Recording. *ASME Journal of Lubrication Technology*, 93(2):272–278, 1972.

UK Data Recovery and HDD repair. http://www.hddtech.co.uk/, March 2008.

Western Digital Corporation. http://www.wdc.com, March 2008.

J. W. White and A. Nigam. A Factored Implicit Scheme for the Numerical Solution of the Reynolds equation at Very Low Spacing. *ASME Journal of Lubrication Technology*, 102:80–85, 1980.

L. Wu. *Physical Modeling and Numerical Simulations of the Slider Air Bearing Problem of Hard disk drives*. PhD thesis, University of California, Berkeley, 2001.

L. Wu and D. B. Bogy. Unstructured Adaptive Triangular Mesh Generation Techniques and Finite Volume Schemes for the Air Bearing Problem in Hard Disk Drives. *ASME Journal of Tribology*, 122:761–770, 2000.

T. Yamada and D. B. Bogy. Load Unload Slider Dynamics in Magnetic Disk Drives. *IEEE Transactions on Magnetics*, 24:2742–2744, 1988.

H. Yamaura and K. Ono. Inverse Analysis of Flying Height Slider Bearings for Magnetic Disk Recording. *Proceedings of the Japan Int. Trib. Conf., Nagoya*, pages 1917–1922, 1990.

Q. Zeng and D. B. Bogy. A Simplified 4-DOF Suspension Model for Dynamic Load/Unload Simulation and its Application. *ASME Journal of Tribology*, 122(1):274279, 2000a.

Q. Zeng and D. B. Bogy. Effects of Suspension Limiters on the Dynamic Load/Unload Process: Numerical Simulation. *IEEE Transactions on Magnetics*, 35(5):2490–2492, 1999a.

Q. Zeng and D. B. Bogy. Slider Air-Bearing Designs for Load/Unload Applications. *IEEE Transactions on Magnetics*, 35(2):746–751, 1999b.

Q. Zeng and D. B. Bogy. Effect of Certain Design Parameters on Load/Unload Performance. *IEEE Transactions on Magnetics*, 36(1):140–147, 2000b.

Q. Zeng and D. B. Bogy. The CML Dynamic Load/Unload Simulator (Version 4.21.40). Technical Report 1999-005, CML, University of California, Berkeley, 1999c.

Q. Zeng and D. B. Bogy. Numerical Simulation of Shock Response of Disk-Suspension-Slider Air Bearing Systems in Hard Disk Drives. Technical Report 2000-003, CML, University of California, Berkeley, 2000c.

Q. Zeng, M. Chapin, and D. B. Bogy. Dynamics of the Unload Process for Negative Pressure Sliders. *IEEE Transactions on Magnetics*, 35:916–920, 1999.

O. C. Zienkiewicz and Y. M. Xie. A Simple Error Estimator and Adaptive Time Stepping

Procedure for Dynamic Analysis. *Earthquake Engineering and Structural Dynamics*, 20:

871–887, 1991.

Appendix A

The CML Dynamic Load/Unload/Shock Simulator (Version 5.1)

A.1 Introduction

This report is a detailed manual for the new Load/Unload and Shock (L/UL/S) simulator developed at the Computer Mechanics Laboratory at the University of California, Berkeley. The response of the suspension-slider-disk system during events such as L/UL and shock is determined by the air bearing behavior as well as the structural response of the suspension and the disk. In this version of the L/UL/S simulator, the air bearing is modeled using the finite volume method as discussed by Hu and Bogy [1995]. The suspension modeling is done using the finite element method with the program taking the pre-assembled mass and stiffness matrices as inputs [see Bhargava and Bogy, 2005].

The new L/UL/S simulator has many improvements over the previous version of
the L/UL simulator [Zeng and Bogy, 1999c]. These include:

1. Finite element modeling for the suspension. The L/UL/S simulator takes as inputs
   pre-assembled mass and stiffness matrices for the suspension. Since a full FE model
   for the suspension would be overkill, we employ a technique of reducing the number
   of degrees of freedom of the system using Guyan reduction [Guyan, 1965] in ANSYS
   [ANSYS, 2005], which is a commercial finite element package.

2. Modeling of user defined ramp profiles, whereby load and unload velocities are calculated automatically using the specified angular motion of the actuator.

3. Simulation of shock and vibration. In addition to simulating the L/UL process, the
   new simulator can also simulate shock and vibration. Shock is modeled as a half-
   sine/square wave acceleration pulse to the system, while vibration is modeled as a
   constant/decaying amplitude sinusoidal acceleration wave.

4. Improved elastic impact modeling. The impact model calculates the force on the
   slider when the fly-height at any point becomes negative. The new model discussed
   in Chapter 3 is more robust and faster than the model used previously.

5. Inclusion of intermolecular forces. The L/UL/S simulator incorporates the intermolecular force model proposed by Gupta and Bogy [2004].

A.2 Installation

The file LULSV51.zip is a compressed zip file of the L/UL/S simulator and the pre

and post processing files. Create the directory CML_DLULS in the root directory of a drive,

eg. c:\CML_DLULS and copy LULSV51.zip into this directory. You can use Winzip to extract

the files from the archive into the CML_DLULS directory, and thereby get the LULSV51.exe

and five subdirectories (mfiles, example1, example2, example3 and example4.

To run a simulation, eg. example1, you should open a new DOS command prompt

(cmd), change the directory to c:\CML_DLULS\example1 (using cd c:\CML_DLULS\example1)

and run the program LULSV51.exe from the parent directory (..\LULSV51.exe). To post

process the results using the provided MATLAB subroutines, you will need to run MAT-

LAB, change the current directory to c:\CML_DLULS\example1 and set path to use the

directory c:\CML_DLULS\mfiles (using path(path, 'c:\CML_DLULS\example1'). Then you

can use the m-files to display the results in MATLAB. eg. force(0,0).

A.3 Procedure

1. Design the air bearing using the CML air bearing design program [Grisso and Bogy, 1999]. Export the rail.dat file for input to the L/UL/S simulator.

2. Create the dynamics.def input file. The easiest way to do this is to use the sample dynamics.def file and modify the desired parameters.

3. Find the steady state flying attitudes of the slider and create the grid. The following procedure suggested by Zeng and Bogy [1999c] is recommended. Use the results from a

static fly-height simulation using the CML air bearing design program as initial values

to run a "no L/UL/S" simulation (L/UL=0, shmod=0 in the *dynamics.def* file) with

adaptive gridding turned on (iadpt=1, ioldgrid=0) and using dt=1e-7, tf=2e-3.

The steady state flying attitudes can be obtained using the post processing programs

to view the attitude histories. The grid created will be saved in the files *x.dat* and

y.dat. Now run the desired simulations with adaptive gridding turned off using the

previously generated grid files.

4. Perform the L/UL or shock/vibration simulations.

5. Analysis of the results can be done using the provided MATLAB postprocessing rou-

tines.

A.4   Suspension model

This section discusses the method for obtaining the mass and stiffness matrices for

the suspension-slider system. The procedure described here uses the finite element model

of the suspension in ANSYS. It is up to the user to export stiffness and mass matrices from

other finite element software. Typically suspension models have a very large number of

degrees of freedom, and solving such large models at each time step for a large number of

time steps is very expensive as well as unnecessary. Hence we reduce the total number of

degrees of freedom of the suspension using a procedure known as Guyan reduction, which

is available in ANSYS as the substructuring option. In the method we discuss here, we will

be able to reduce the suspension model as well as obtain the mass and stiffness matrices in one single step. Since the suspension model is inherently nonlinear, due to contacts at the dimple-flexure and the limiters, we will obtain the mass and stiffness matrices for the free state (i.e. when the dimple is open and limiters are not in contact) of the suspension only, and implement the contacts on top of the matrices using additional contact elements in the L/UL/S simulator itself. All contacts are modeled as node to node contacts. The following procedure discusses how to reduce the finite element model and obtain the mass and stiffness matrices:

1. Open the suspension model (*.db, *.inp etc) in ANSYS. Enter the solution mode.

2. Unselect all contact elements

3. Identify node numbers corresponding to our primary DOF of interest. These are:

    (a) The six DOF for the attitude of the slider (3 displacements and 3 rotations, see Figure A.3)

    (b) DOFs for the dimple-flexure, and limiters which will be used for contact elements

    (c) DOF for the L/UL tab. This will be used to implement contact with the ramp, and can also be used to monitor the motion of the load beam during the event of shock.

4. Define the DOFs identified above as master DOFs using the M command.

5. Define the total number of master DOFs using the TOTAL command (a total of 250 MDOFs are recommended, i.e. TOTAL,250).

6. Define a new substructuring analysis. Set the analysis options to generate Stiffness
   and Mass matrices along with load vector and matrix printouts enabled.

7. Redirect output from the screen to a file *SuperNNN.txt* using the /OUTPUT command,
   where NNN is the total number of master DOFs defined using the TOTAL command.

8. Run the substructuring analysis

9. After the analysis save the listing of the master DOF set using the NSEL,S,M,,ALL and
   NLIST commands. Save the output from the NLIST command to a file *coordsNNN.txt*.

10. Now use the fileprocessor program using the command *fileproc.exe NNN* in the same
    directory as the files *superNNN.txt* and *coordsNNN.txt* to generate the files *mass.txt*,
    stiffness.txt and *coords.txt*.

The generated stiffness, mass and coordinate data files can be used as inputs to the L/UL/S
simulator.

A.5   Input files

A.5.1   rail.dat

The rail.dat file defines the rail shape and the air bearing surface. This file is
generated by the CML Air Bearing Design Program [Grisso and Bogy, 1999].

A.5.2   dynamics.def

The dynamics.def file is the primary parameter input file to the L/UL/S simulator. Most of them are the same as described in Chen et al. [1998]. The parameters which are important, different or additional are discussed here.

Problem definition

f0: suspension normal load (kg)

xf0(/xl), yf0(/yl): normalized coordinates of the load point. They must be same as xg and yg.

rpm: disk revolutions per minute

dt: time step (s). A value of 1e-7 is recommended for all simulations.

ra: radius of the *reference* position (see Figure A.1).

Suspension

iact: 0 = no actuator; 1 = inline actuator. Always keep iact = 1.

xact: length of actuator arm [m] (see Figure A.1).

dact: angular position of actuator with respect to the reference position [rad] (see Figure A.1).

vact: angular velocity of actuator, positive velocity indicates that slider moves from OD to ID [rad/s].

`ske`: the skew angle (degree) at the specified reference radial position (see Figure A.1).

Initial Flying Condition

`hm, hp, hr, dx, dy and yaw`: the initial nominal flying height (m), pitch, roll (rad), x, y displacements (see Figure A.3) and yaw (rad) of the slider. For loading, these values are not used.

Grid Control

`iadpt`: 1 = use adaptive grid to generate the grid; 0 = disable grid generation.

`ioldgrid`: 1 = use the old grid saved in the `x.dat` and `y.dat` files; 0 = use adaptive grid.

`nx, ny`: grid size in the x and y directions, respectively. Must be in the form of $16 \times n + 2$. Usually, these should be larger than 146.

`difmax, decay`: used in the adaptive grid. See Zeng and Bogy [1999c] for more information.

Asperity Contact

`icmod`: Asperity contact model: 1=GW model; 2=elastic-plastic model. We suggest using the GW model here.

`gldht`: glide height (m).

Dynamic Load/Unload

L/UL: 0=disable L/UL simulation; 1=simulate the load process; 2=simulate the unload process.

OutPr5: 0=no pressure profile output; 1=output pressure profiles at the specified times.

p@t1(ms), p@t2, p@t3, p@t4: output pressure profiles at these times [ms] if OutPr5 is equal to 1.

S.pitch, S.roll: static pitch (PSA) and roll (RSA) [rad].

I_ey, I_ydst: composite elastic modulus and yield strength [Pa] used in calculating slider/disk impact if the clearance between the slider and disk at some point is less than zero.

suspsz: Total number of DOFs in the suspension

dofux, ..., dofrotz: DOF numbers corresponding to slider x, y, z displacements and x, y, z rotations (roll, pitch and yaw).

doftab: DOF number corresponding to L/UL tab z-displacement.

dtfac: Time step reduction factor during contact. It is recommended this value be kept at 10 (i.e. the time step will be reduced by a factor of 10, when contact occurs). For quicker simulations a value of 1 may be used with some loss of accuracy during contact.

nocele: Number of contact elements to be defined. *nocele = # of limiter contacts+*
1 (for dimple).

dofcu, dofcd, constat, preload: DOF number corresponding to nodes above
(dofcu) and below (dofcd) contact element. constat indicates whether the contact elements are closed (1) or open (0) at the beginning of the simulation. preload
indicates the preloading of the contact element. This can be used to adjust the dimple
preload [mN] for the dimple contact element (otherwise leave at 0).

nrp, theta, z: The description of the ramp profile. nrp indicates the number of
ramp profile points. The ramp profile will be linearly interpolated between these
points. theta indicates the angle [rad] measured from the reference position and z
indicates the height of the ramp [mm] (see Figure A.4 and Figure A.5).

Disk Modeling

idmod: 1 is disk modeling on and 0 indicates disk is not to be modeled.

disksize: Size of the disk model. The size of the default disk model is 247.

nsnodes: Number of radial nodal locations.

zdof, rotx, roty: DOF numbers for z-displacement and x,y rotations for nodes
from the OD to the ID on the x-axis. A second order linear interpolant will be used
to calculate disk motion between these radial locations.

Shock

shmod: Shock mode: 0 noshock, 1 half-sine shock pulse, 2 square-wave shock pulse, 3 constant magnitude sinusoidal acceleration field, 4 decaying sinusoidal acceleration field.

sttime: Acceleration pulse start time [s]

pulsewid: Pulse width of the acceleration pulse/wave [s].

magnitude: Magnitude of acceleration [G].

tconst: Time constant for shmod = 4.

A.5.3 mass.txt, stiffness.txt, coords.txt

The first two files are the mass and stiffness matrices for the suspension. The third file contains the coordinate locations for the nodes as well as DOF information. The columns in *coords.txt* are x, y and z coordinates of the DOFs, and the type of DOF (1,2,3: x,y,z displacement; 4,5,6: x,y,z rotation, see see Figure A.2). These files can be automatically generated from ANSYS using the procedure in the previous section. The user can also choose to generate these files using other finite element packages.

A.6   Output files

This section discusses additional time-history output files generated by the L/UL/S simulator.

A.6.1   attitude.dat

This file contains the time [s], the displacements of the 6 DOF of the slider [mm, rad], as well as velocities [mm/s, rad/s]and accelerations [mm/s/s, rad/s/s] corresponding to these 6 DOF (see Figure A.3).

A.6.2   cestatii.dat

Time [s], and the contact status, contact spacings [mm] and contact force [mN] for each contact element.

A.6.3   cpressures.dat

Time [s], maximum contact pressure [atm] and maximum impact pressure [atm].

A.6.4   imf.dat

Time [s], intermolecular force [mN], resultant intermolecular moments in pitch and roll directions about slider center [mN mm].

A.6.5   lultab.dat

Time [s], ramp contact status, tab displacement [mm], ramp contact force [mN], angle dact [rad] (see Figure A.1) and the ramp height at dact [mm].

A.6.6   shock.dat

Time [s], acceleration magnitude, disk z-displacement [mm], disk 'roll' [rad] and disk 'pitch' [rad].

A.7   Post-processing

All of the post processing programs available for use with the previous version of the CML L/UL Simulator [Zeng and Bogy, 1999c] can still be used with the L/UL/S Simulator. However some of the old Matlab subroutines have been modified and certain new ones have been added, which are discussed in this section.

A.7.1   displaceLUL.m

Usage: displaceLUL(<n1>,<n2>,<'Comment1'>,<'Comment2'>,<'Comment3'>,<'Comment4'>) where <n1> and <n2> indicate the range of the data to be plotted. Setting <n1> and <n2> will plot the entire time history data.

This function was previously available as *displace.m* in the previous version of the L/UL simulator. However it can still be used to plot L/UL displacements. The output from the command displaceLUL(0,0,'Loading Process','Vlul = 8 rad/s','',''') for example 1 is shown in Figure A.6. The various quantities plotted are: a) absolute displacements of L/UL tab and slider center, b) the nominal FH, c) the minimum clearance, d) the pitch and e) the roll.

A.7.2   displaceSHOCK.m

Usage: displaceSHOCK(<n1>,<n2> where <n1> and <n2> indicate the range of the data to be plotted. Setting <n1> and <n2> will plot the entire time history data.

A modification of the function *displaceLUL.m*, this function can be used to plot the slider attitude for shock and vibration simulations. The output from the command `displaceSHOCK(0,0)` for example 3 is shown in Figure A.7 and for example 4 in Figure A.8. The various quantities plotted are: a) the acceleration profile, b) absolute displacements of load beam, slider center (flexure) and disk, c) the nominal FH, d) the minimum clearance, e) the pitch and f) the roll.

A.7.3 force.m

Usage: `force(<n1>,<n2>)` where `<n1>` and `<n2>` indicate the range of the data to be plotted.

This function has been retained from the previous version of the L/UL simulator. The output from the command `force(0,0)` for example 2 has been plotted in Figure A.9 and for example 3 in Figure A.10. The various quantities plotted are: a) the air bearing forces (positive, negative and net), b) the bearing force center, c) the asperity contact forces and d) the elastic impact forces.

A.7.4 cestat.m

Usage: `cestat(<n1>,<n2>)` where `<n1>` and `<n2>` indicate the range of the data to be plotted.

This function plots the spacing and contact forces corresponding to the contact elements defined for the suspension. These are the dimple-flexure and the limiters. Figure A.11 plots the contact element status for example 2 (unloading) generated using the command `cestat(0,0)`, and Figure A.12 plots for example 3. The various quantities plotted are: a) dimple spacing, b) dimple contact force, c) limiter spacing and d) the limiter contact forces.

A.7.5   lulbehav.m

Usage: `lulbehav(<n1>,<n2>)` where `<n1>` and `<n2>` indicate the range of the data to be plotted.

This function plots various quantities relating to the L/UL behavior. The output from the command `lulbehav(0,0)` for example 1 have been plotted in Figure A.13. The quantities plotted are: a), b) absolute displacements of the ramp (profile under the L/Ul tab) and the L/UL tab as a function of time and actuator angle, c) the ramp contact force, d) angular displacement of the actuator, e) velocity of the L/UL tab and f) the acceleration of the L/UL tab.

A.7.6   actmot.m

Usage: `actmot(<n1>,<n2>)` where `<n1>` and `<n2>` indicate the range of the data to be plotted.

This function plots various quantities relating to the actuator motion. The output from the command `actmot(0,0)` for example 1 have been plotted in Figure A.14. The quantities plotted are: a), b) absolute displacements of the ramp (profile under the L/Ul tab) and the L/UL tab as a function of time and actuator angle, c) the ramp contact force, d) angular displacement of the actuator, e) velocity of the L/UL tab and f) the acceleration of the L/UL tab.

A.8   Figures

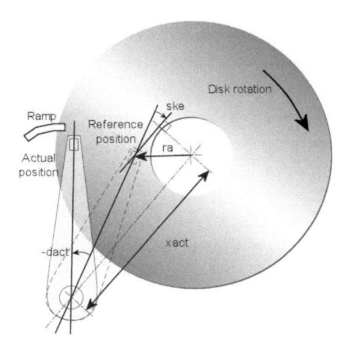

Figure A.1: Actuator nomenclature

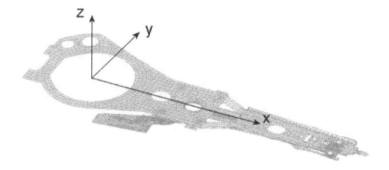

Figure A.2: The suspension coordinate system

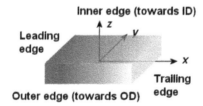

Figure A.3: The slider coordinate system

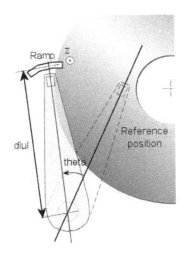

Figure A.4: The ramp coordinates

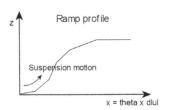

Figure A.5: Sample ramp profile

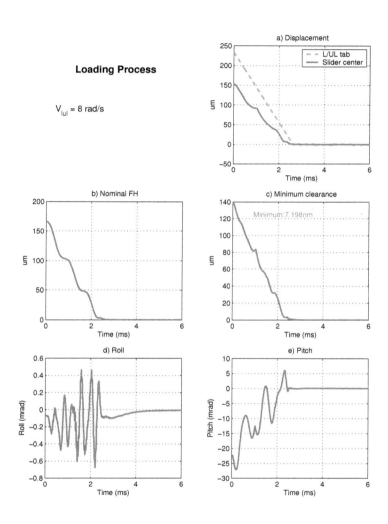

Figure A.6: Slider attitude for example 1 plotted using the *displaceLUL* command

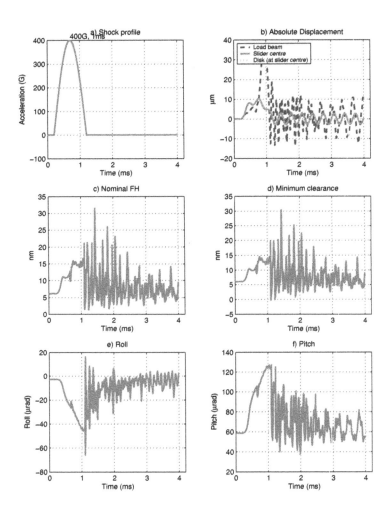

Figure A.7: Slider attitude for example 3 plotted using the *displaceSHOCK* command

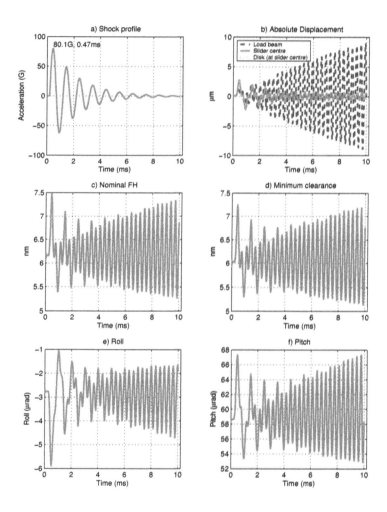

Figure A.8: Slider attitude for example 4 plotted using the *displaceSHOCK* command

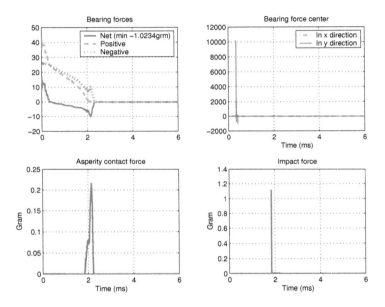

Figure A.9: Air bearing and contact forces for example 2 plotted using the *force* command

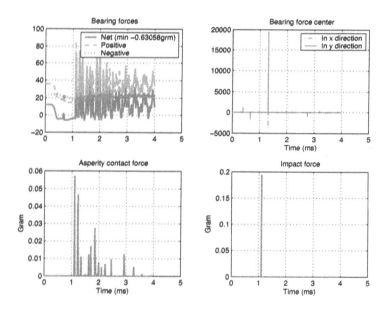

Figure A.10: Air bearing and contact forces for example 3 plotted using the *force* command

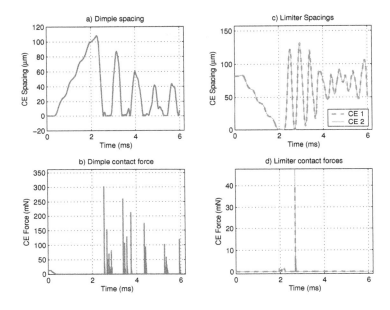

Figure A.11: Contact element data for example 2 plotted using the *cestat* command

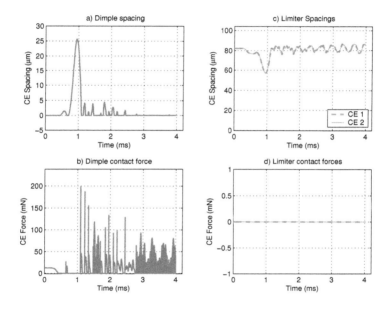

Figure A.12: Contact element data for example 3 plotted using the *cestat* command

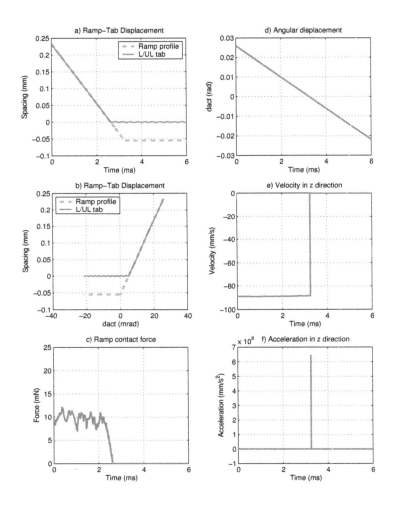

Figure A.13: L/UL tab behavior for example 1 plotted using the *lulbehav* command

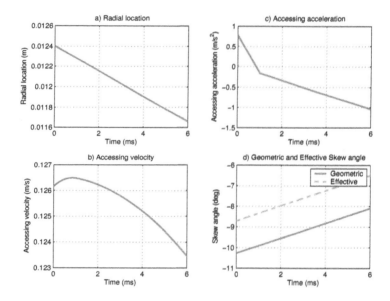

Figure A.14: Actuator motion for example 1 plotted using the *actmot* command

www.ingramcontent.com/pod-product-compliance
Lightning Source LLC
LaVergne TN
LVHW042332060326
832902LV00006B/130